UK RESEARCH IN ADVANCED MANUFACTURE

IMechE CONFERENCE PUBLICATIONS 1986–12

Sponsored by
The Manufacturing Industries Division of
The Institution of Mechanical Engineers

Co-sponsored by
The Application of Computer and Manufacturing Engineering
Directorate of the Science and Engineering Research Council
The Institution of Electrical Engineers
The Institution of Production Engineers

10–11 December 1986

The Institution of Mechanical Engineers
Birdcage Walk
London

Published for
The Institution of Mechanical Engineers
by Mechanical Engineering Publications Limited
London

Proceedings of the Institution of Mechanical Engineers

621.78

UK

ML

UK RESEARCH IN
ADVANCED MANUFACTURE

This book is
the last

Conference Planning Panel

T Husband, PhD, MA, CEng, FIMechE, FIProdE (Chairman)
Department of Mechanical Engineering
Imperial College of Science and Technology
London

C Andrew, PhD, MA, CEng, FIMechE
Engineering Department
Cambridge University
Cambridgeshire

J Billingsley, PhD, MA, CEng, FIEE, FIOA
Portsmouth Polytechnic
Hampshire

B Davies, MPhil, CEng, MIMechE
Department of Mechanical Engineering
Imperial College of Science and Technology
London

P J Drazan, PhD, MSc, CEng, MIMechE
Department of Mechanical and Manufacturing
   Systems Engineering
University of Wales Institute of Science and Technology
Cardiff

P Smith, BSc
Science and Engineering Research Council
Swindon
Wiltshire

CONTENTS

The Institution of Mechanical Engineers

The primary purpose of the 76,000-member Institution of Mechanical Engineers, formed in 1847, has always been and remains the promotion of standards of excellence in British mechanical engineering and a high level of professional development, competence and conduct among aspiring and practising members. Membership of IMechE is highly regarded by employers, both within the UK and overseas, who recognise that its carefully monitored academic training and responsibility standards are second to none. Indeed they offer incontrovertible evidence of a sound formation and continuing development in career progression.

In pursuit of its aim of attracting suitably qualified youngsters into the profession — in adequate numbers to meet the country's future needs — and of assisting established Chartered Mechanical Engineers to update their knowledge of technological developments — in areas such as CADCAM, robotics and FMS, for example — the IMechE offers a comprehensive range of services and activities. Among these, to name but a few, are symposia, courses, conferences, lectures, competitions, surveys, publications, awards and prizes. A Library containing 150,000 books and periodicals and an Information Service which uses a computer terminal linked to databases in Europe and the USA are among the facilities provided by the Institution.

If you wish to know more about the membership requirements or about the Institution's activities listed above — or have a friend or relative who might be interested — telephone or write to IMechE in the first instance and ask for a copy of our colour 'at a glance' leaflet. This provides fuller details and the contact points — both at the London HQ and IMechE's Bury St Edmunds office — for various aspects of the organisation's operation. Specifically it contains a tear-off slip through which more information on any of the membership grades (Student, Graduate, Associate Member, Member and Fellow) may be obtained.

Corporate members of the Institution are able to use the coveted letters 'CEng, MIMechE' or 'CEng, FIMechE' after their name, designations instantly recognised by, and highly acceptable to, employers in the field of engineering. There is no way other than by membership through which they can be obtained!

C376/86

The implementation of robotics in a food manufacturing company

P R DRAYSON, BSc, PhD
Trebor Limited and Aston University
J FLECK, MA, BSc and K FOSTER, MA, PhD, CEng, FIMechE
Department of Mechanical Engineering, Aston University

SYNOPSIS Action-research of the method by which a large food company investigated, evaluated, and adopted robot technology is reported. It was found that many applications were not cost effective and required excessive floorspace in comparison with conventional manufacturing methods. Integration of the robot systems and organisational resistance to change were also significant factors. Outline specifications for robot systems to meet the food industry's needs were derived, an alternative method of financial appraisal employed, and a new parameter for assessing robot system design developed. It is concluded that an integrated approach to managing the adoption process within a longer-term planning orientation is crucial to successful robot use.

1 INTRODUCTION. THE PROCESS OF ROBOT ADOPTION.

The implementation of any new manufacturing technology by a company is a complex task (1). Indeed it has been argued that the problems associated with A.M.T. adoption and use are due to the technology being too far ahead of our ability to manage it (2,3). These problems are reflected in the poor rates of UK robot adoption in particular. Research at Aston on the diffusion of robotics identified some of the reasons behind the UK's poor performance in adopting robot technology (4) and the complexity of the technical, financial and managerial factors inherent in robot adoption at the level of the firm (5).

This paper reports upon a research study (6) on the adoption of robotics by Trebor Limited. In the food industry robots are not yet standard and the use of this technology was a radical new departure for the firm, constituting a case of significant manufacturing innovation. The study was carried out by a direct participant in the work and thus provided a novel perspective of robot adoption in action. Our over-riding conclusion is that an integrated approach within a longterm planning orientation is crucial to successful robot use. The implementation of individual robot applications must be regarded as components in a continuous process of adoption involving consideration of the following points:

* the method used to identify potential robot applications and to design successful systems;

* the financial appraisal of the viability of the investment taking account of the company's business strategy;

* the development of in-house expertise in the application of robotics to the company's own particular manufacturing situation;

* the management of what is an innovative project within the established structure of the organisation;

* the specification of the component robot-technology required to meet the company's needs.

This paper develops this conclusion by discussion of an example application in the context of the overall robot adoption process. Fuller details and background information on this research are provided elsewhere (6).

This process of robot adoption was found to conform in general terms to the model of innovation proposed by Rogers and Shoemaker (7), in that it comprised various stages: interest and investigation of robotics, evaluation of the potential for the company and trial adoption.

These stages are therefore used as a framework for the following discussion.

2. INTEREST AND INVESTIGATION

Interest in robotics first arose from the wish to further apply microprocessor technology to batch and continuous-process manufacturing. This interest was sharpened by the consideration of robot technology by a key competitor and led to the start of a project aimed at assessing the potential and implications of the technology for the company. The first stage of this project was a survey of the manufacturing processes at each factory which identified a total of twenty-eight potential robot applications, see Table 1 below.

By type of application:

	No.	%
Packaging	12	34
Palletising	10	29
Materials Handling	10	29
Assembly	3	8
	35*	100

By type of product:

	No.	%
Major brand	11	39
Brand pack derivative	5	18
	16	57
Wrapped weighout	2	7
Unwrapped weighout	1	4
Children's line	7	25
Total commodities	10	36
Others	2	3
Total	28	100

* Double counting for multiple-operation applications.

Table 1 Summary of the potential robot applications identified in Trebor.

3. EVALUATION

Following the survey the applications were then analysed in terms of four parameters: payload, cycle-time, horizontal reach and vertical reach, see Figures 1, 2 and 3.

It was found from this analysis of the data that:

a) cycle-times were generally less than 10s, with a large proportion in the 1.5-3s band. These are very low for robot operations and suggest operating speeds 100-300 per cent faster than those currently available;

b) payload fell into two main groups: 10-25 kg and 0-4 kg. Although these figures are well within current capabilities they are not available with the speed/reach capabilities required;

c) horizontal and vertical reach also fell into two groups:

   0.1-0.2m vertical reach
   R0.7 - R1.0m horizontal

   1.6 - 2.0m vertical reach
   R2.0 - R2.5m horizontal

These results indicated that two types of robot were in fact necessary to meet the company requirements, with the following specifications

	Type A for small items	Type B for large items
Max load/kg	4	25
Horizontal reach/m	R1	R2.5
Vertical reach/m	0.2	2
Maximum speed/ms^{-1}	3	2

Following this investigation and evaluation of the potential for using robots, further more detailed studies of selected projects were carried out. These involved the preparation of design solutions and the eventual implementation of one of the proposals as a trial adoption of the technology.

4 TRIAL ADOPTION

The design and implementation of the first installation was found to be a highly iterative process; several applications had to be considered in parallel in order to ensure that, in at least one case, solutions were found to the diverse technical and non-technical problems which were encountered, see Table 2 below.

Disadvantage of the robot solution	No of applications
Required excessive floorspace	15
Payback exceeded 3 years	11
Complex integration problems	6

Table 2 The three most common problems encountered in the Trebor Project

If we look then at the considerations involved in implementing a trial adoption of robots within the company we find that three issues predominate: floorspace, payback and system integration.

These points are further discussed below with reference to an example robot-application common to many fast-moving-consumer-goods manufacturers: end of line packaging and palletising. At present these tasks are generally done by hand where the volume of output is too low to justify investment in automatic machinery. However, flexible automation offers the possibility of integrating several of these low-volume lines at the packaging/palletising operation - providing significant operational and cost benefits.

4.1 Floorspace

The floorspace needed by a robot-based manufacturing cell is related to the robot working area requirement, which is a function of both the task to be carried out and the configuration of the robot manipulator. Packaging and palletising operations require most working area in the horizontal plane as illustrated by Figure 3.

These areas are the minimum required, specified by the size of the objects which the robot is working on, in this case small cartons and palletised loads. To meet these minimum dimensions a robot may have to be used with a larger reach than is needed to meet some other parameter, such as load capacity, or because the robot work-envelope is itself shaped inefficiently.

For example robots of revolute configuration providing 25 kg load capacity will typically require a minimum of 35m² floor area (horizontal reach enveloped plus 0.5m guarding allowance) but only provide a work area of 3.5m² at a vertical reach of 2m as required for palletising – a usable work-space which is only 10 per cent of the floor area it occupies, comparing poorly with the existing manual method. This result is surprising because both the literature and the robot manufacturers cite compactness and space efficiency as major advantages of robot technology.

This suggests a new criterion of PRODUCTIVE WORKSPACE (P); where P equals the percentage of the total floor area taken up by the cell that is available for use at the working height specified by the application.

The notion of productive workspace was used to assess various design solutions based on different robot models. It was found that robots of gantry configuration produced the highest values of P in this type of application - typically greater than 60 per cent. This simple parameter was of considerable use in designing more space efficient systems.

4.2   Financial appraisal

In parallel with the analysis and design work involved in developing a technically viable system, financial analysis of the return on the capital investment required is also essential. The appraisal technique used at Trebor, as well as the majority of UK companies (8), is that of payback.

Payback appraisal for a typical palletising application would be as follows:

Capital	£
Robot	35 000
Tooling etc	40 000
M/c modifications	20 500
	95 500

Savings per year

| Direct Labour | 13 000 |

$$\text{Payback} = \frac{£95\ 500}{£13\ 000} = 7\ \text{years}$$

This figure of seven years is considerably in excess of the three-year payback criterion applied by many companies

However, it was recognised in our case that there were additional factors relevant to the investment decision which were not included in the payback appraisal. These included: an improvement in the quality of work (the manual task was heavy, repetitive and disliked by operators); an improvement in the consistency of packaging (improved quality and reduced product damage during distribution). Managers recognised these factors as important but the standard investment appraisal technique distracted them from considering all aspects of the decision by quantifying only some of the factors that had to be considered. To overcome this, a very simplified form of cost benefit analysis was used to quantify and make explicit managers' implicit

assessments of the relative values of the costs and benefits relevant to the application under consideration.

The projects were assessed using an appraisal sheet which listed the possible costs and savings in accordance with the Company's Production and Personnel policies. Each item, both quantifiable and non-quantifiable in cash terms was then rated on a scale of 1 to 10 by the team carrying out the appraisal. This rating was a measure of their assessment of the relative importance of this benefit or cost to the overall viability of the project. (i.e. 1 : not at all important to 10: the most important cost or benefit). Next, the items which were quantifiable in cash terms, for example direct labour cost, were used to derive an equivalent cash value for all the indirect, qualitative or intangible factors, in proportion to the rating assessment.

In the cost of the palletising project for example, the new appraisal might be as follows:

ITEM	RATING	REAL CASH £	EQUIVALENT VALUE £
Costs (investment)			
Robot	7	35 000	
Tooling etc M/C	8	40 000	
Modifications	4	20 500	
Disruption	5		25 000
Sub-Total	24	95 500	25 000
Benefits (per year)			
Direct labour cost	4	13 000	
Health and safety	8		26 000
Quality	6		19 500
Sub-total	18	13 000	45 500

$$\text{Payback} = \frac{(\text{real cash} + \text{equivalent value}) \text{ cost}}{(\text{real cash} + \text{equivalent value}) \text{ benefits}}$$

$$= \frac{£95\ 500 + £EQ\ 25\ 000}{£13\ 000 + £EQ\ 45\ 500}$$

$$= 2\ \text{years}$$

Thus the payback figure included a measure of all relevant factors in a clear and concise way, which made explicit the managers' own assessments of the importance of the normally unquantified elements. This technique is one way in which intangible factors in the analysis may be made more concrete and easier to compare, without complex, detailed investigations which in reality companies are unlikely to use.

Figure 4 illustrates the results of applying this technique to nine applications which were studied in detail. Not only did this produce an overall improvement in pay-back across the board, but much more significantly, the order of projects was changed dramatically. This suggests that selection of the best application for the firm may depend upon explicit consideration of the broader issues.

4.3   System Integration

For the overall manufacturing process to work well, the equipment and people within it must function effectively together as well as individually. Achieving this successful integration of the robot palletising cell was complicated by a number of factors.

Firstly, the existing equipment which was to be interfaced with the robot cell was not originally designed with this in mind. For example, orientation was not maintained once the product had left the outfeed of upstream wrapping machines. Secondly, the process was quite labour intensive with nearby operators available to 'nurse' problematic equipment, to allow for variations in the process and to carry out subtle, but often very complex tasks such as product inspection. Thirdly, packaging machines are generally designed as discrete, autonomous units which require little interfacing to the existing process. The company has been able to take a localised approach to new machine installation and had little experience in overcoming machine integration problems.

The operation of the robot cell therefore required significant changes to the existing process - both in terms of equipment modifications and working practices. These changes would also represent a significant proportion of the capital cost of the system; in the case of end of line packaging and palletising it was found that these integration costs represented as much as two-thirds of total system cost - substantially higher than the 50% proportion often reported in other industries, clearly showing the link between the financial viability of the application and the level of peripheral equipment needed to interface the robot system to the existing manufacturing process. On the other hand, applications proposed as part of a complete new process installation were more cost-effective because a total-system perspective avoided such integration problems. In the case of retro-fitted systems, having a long term plan for the eventual modernisation of the overall manufacturing process concerned allowed the adoption of a top-down approach to system-design which provided a similar cost advantage.

Thus far we have focussed on the considerations arising during the implementation of the first trial application. As was pointed out at the start of this paper however, the introduction of industrial robots is a continuous process which also depends upon the development of in-house skills and the management of what in fact constitutes an innovative project within the company's normal operations.

5   GENERAL MANAGEMENT OF THE PROCESS OF ROBOT ADOPTION

It has been shown above how important it is that the robot system is effectively integrated into the existing manufacturing process. For this to happen two bodies of expertise have to be brought together.

a)   the knowledge and experience embodied in the people running the existing process and;

b)   the skill in application engineering held by the robotics specialists.

A synthesis of this knowledge was found to be essential for the development of successful system solutions (9). However, for this synthesis to happen there has to be commitment from both sides, whereas for the managers and engineers responsible for a particular area the introduction of a robot system was often not in their immediate interest; in the short-term it would be likely to cause some disruption of production which would increase their workload or be an unwelcome extra burden. Without their positive commitment, this crucial synthesis of robot and local knowledge was lacking and technically viable systems could not be developed which overcame the problems discussed earlier.

The key to gaining this joint commitment was again found to depend upon taking a broad, long-term view of introducing robots as a continuous process. A long-term plan setting out why adopting robotics was important to the future competitive position of the company as a whole, put these short-term difficulties at the local level into perspective - giving people clear, jointly-shared aims which overcame their personal concerns. At the operational level, the difficulties inherent in individual applications may not seem justified - whereas in the long-term the capabilities offered by the technology may be vital to the company's success. It is here that the financial dimension of the adoption process becomes so critical, because if, as we have shown, the introduction of robots is not a short-term operational decision but a strategic one, then localised assessments of viability using payback techniques for example, are wholly inappropriate. The financial appraisal should not be carried out on individual, discrete projects but on the overall competitive advantage offered by the long-term introduction of flexible automation.

However, identifying whether this competitive opportunity exists or not is a difficult task in itself. Another problem is the shortage of individuals who have the specialist expertise in robotics together with the breadth of scope needed to assess the technology in the context of the company's business strategy and competitive position. Again the Trebor experience suggests the answer to this problem lies in the manner by which the adoption process is managed.

Extensive previous research has shown that for an idea to be taken up and exploited, by a company for example, there has to be one or more people 'behind it' - positively going out of their way to promote and champion the idea within the company because they believe in it so strongly (10).

This is essentially a political process, requiring sensitivity to a wide range of issues and the willingness to campaign over what may be a substantial length of time. It is precisely this kind of approach which is needed by the robot adoption process discussed above, especially in firms where the adoption of robots is a novelty. This suggests that if companies can identify natural robot champions within their organisation, enable them to acquire the appropriate training and can encourage them to take this broad, integrated approach then they will considerably increase the probability of a successful introduction of the technology.

6. CONCLUSIONS

Thus although the successful introduction of industrial robot technology is a difficult and complex task, it can be achieved by means of a multi-disciplinary approach.

Specifically, a longer term planning orientation will reduce both implementation costs and organisational resistance to change as well as providing a balanced framework for financial evaluation. Equally, taking an integrated approach to the adoption process as a whole provides a common philosophy for the design method, the project management method and the financial appraisal method.

REFERENCES

1. BESSANT, J., Influential factors in manufacturing innovation, Research Policy (II), 1982, pp. 117-132.

2. HUSBAND, T.M., The Impact of New Technology on Production, Omega, Vol.15. No.3. 1984.

3. Advisory Council for Applied Research and Development. New Opportunities in Manufacturing. The Management of Technology London: HMSO. October 1983, p.30

4. ZERMENO-GONZALEZ, R., The Development and Diffusion of Industrial Robots, PhD thesis, Aston University, 1980.

5. FLECK, J., The Introduction of the Industrial Robot in Britain, Robotica, Vol.2, 1984, pp.169-175.

6. DRAYSON, P.R., The Implementation of Industrial Robots in a Manufacturing Organisation, PhD Thesis, Aston University, 1986.

7. ROGERS, E.M., and SHOEMAKER, F.F., Communication of Innovations (2nd edition), New York: The Free Press, 1971.

8. PIKE, R.H., Capital Budgeting in the 1980's London: I.C.M.A., 1982.

9. FLECK, J., The Effective Utilisation of Robots: The Management of Expertise and Knowledge, Proc. 2nd Conf. on Automated Manufacturing, Bedford: IFS, 1983.

10. SCHON, D.A., Champions for Radical New Inventions, in Hainer et al (eds) Uncertainty in Research Management and New Product Development, Reinhold, 1967.

ACKNOWLEDGEMENTS

The authors gratefully acknowledge the assistance of R W Clack, BSc and A. Gregory, MIMechE of Trebor Limited during this research, and wish to thank the Science and Engineering Research Council and Trebor Limited for their financial support and for their permission to publish this work.

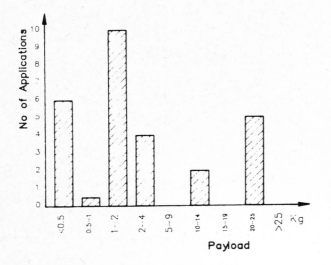

Fig 1   Analysis of applications by payload

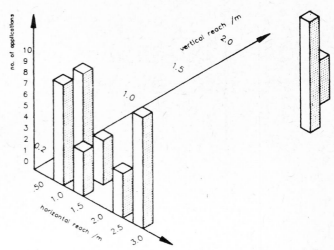

Fig 3   Analysis by horizontal and vertical reach

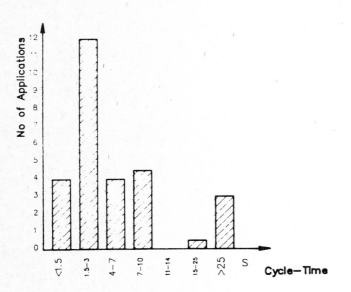

Fig 2   Analysis by cycle-time

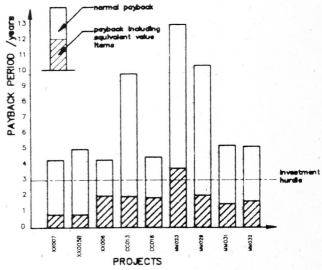

Fig 4   Graph of real and equivalent payback period

C374/86

Implementation of a computer aided manufacturing system

T W STACEY, BSc and **A E MIDDLEDITCH**, BSc(Tech), MSc, MS, PhD
School of Engineering and Science, Polytechnic of Central London

SYNOPSIS The data necessary to define material removal manufacturing processes at the level of volumes and operations thereon is described. Such definitions may be considered as process plans or high level part programs. They provide a useful specification of the interface between process planning systems and tool path planning and generation systems. The manufacturing definitions are heirarchically subdivided. This simplifies understanding, suggests a data structure implementation and allows software to be developed with significant information hiding properties. It also suggests a block structural language for process plans.

The proposed manufacturing definitions are used by geometric algorithms for data compatibility, (i.e. process plan validity) checks, suitable tool path existence checks, and tool path generation. A prototype implementatiuon of a computer aided manufacturing system based on these algorithms is described. Although the theoretical framework is of general applicability, this prototype is limited to 2.5D components machined on 3 axis milling machines. Manufacturing definitions for a typical component and the NC commands generated by the system are described. Associated computer generated images of these components are shown for comparison with the actual component machined on a machining centre. These images are generated from volume models which form part of the manufacturing process definition.

1.0 INTRODUCTION

This paper describes a prototype implementation of a computer—aided manufacturing system. This CAM system integrates a previously proposed definition of material removing manufacturing processes [Bailey & Middleditch 1978] with purely geometric methods and algorithms for checking the data and generating machine tool commands [Stacey & Middleditch 1986]. Manufacturing process definitions may be obtained from user supplied or automatically generated process plans. Since the prototype implementation uses the former approach for simplicity, automatic process planning is outside the scope of this paper.

Data checking and machine tool command generation rely extensively on geometric utilities for the manipulatiuon of two and three dimensional point sets (areas and volumes). However, detailed discussion of such utilities is also outside the scope of this paper. Machine tool commands are translated into a machine tool controller readable form via a device independent GKS—like procedural interface.

1.1 System Structure

The system is sub—divided into separate modules on the following functional basis:

1. Input/output of manufacturing data.
2. Manufacturing data structure management.
3. Data compatibility checking.
4. Machine tool command generation.

Figure 1 illustrates the software heirarchy, while figure 2 illustrates information flow within the system. This paper is concerned with those elements of the system enclosed by the dashed box. Since external facilities such as the user interface and graphical and geometric operations are not unique to a CAM system, they are not discussed further.

The data management module maintains the data structures for manufacturing definitions. Modules above it in figure 2 are concerned with creating manufacturing definitions, and modules below it are concerned with their execution. Data compatibility checking may be regarded as trial execution.

The generation of manufacturing process plans is concerned with such technological issues as the selection of different setups (workpiece location, clamping arrangements etc.), cutters, cutting conditions or even different machine tools for the completion of a particular component. The necessary technological information is supplied by the operator in the prototype implementation.

The data structure for manufacturing process definitions is reviewed in section 2 and methods for obtaining the necessary data are discussed in section 4. Some modifications to the data structure previously proposed [Bailey & Middleditch 1978] are introduced for the current implementation. Geometric methods for checking data and generating machine tool commands are reviewed in sections 3 and 5.

2.0 DATA STRUCTURE

For reasons of clarity and efficiency, the representation of data (both geometric and technological) necessary to define a machining operation should reflect the manner in which a machine tool operates. This may be achieved by grouping data into independent units arranged in a non—binary heirarchy. A five level heirarchy is described in [Bailey & Middleditch 1978]. However, it is preferable to add a further level in order to distinguish reorientation of an existing workpiece from change of workpiece (fig. 3):

1. **Workpiece** : Geometric specification of the initial workpiece.

2. **Setup** : Workpiece location and geometric specification of jigs, fixtures and any other collision avoidance regions associated with the setup.

3. **Cutter** : Geometric specification of the cutter, including cutter shank, cutting volume (that part of the cutter which is capable of material removal) and information about preferred and allowed directions of cutting.

4. **Process Parameters** : Machining process parameters including feed—rate, spindle—speed, coolants, maximum depth and width of cut. (A more advanced implementation should derive this data from cutter and workpiece material properties).

5. **Cavity** : Geometric specification of the volume to be removed.

6. **Location** : Position and orientation of a a specific cavity.

The data at any node on the tree is relevant at all nodes of its subtrees. Thus, the path from leaf to root contains all the data necessary to machine a particular instance of a given cavity. Conversely, a single setup could involve several cutters, each operating under several sets of cutting conditions, on each of several cavity shapes, each cavity at one or more locations.

The restricted domain of a finite number of semantically distinct levels, with the associated separation of logically distinct units of data simplifies data structure interrogation. This practice is not necessarily followed in conventional part programming languages. For example, the POCKET statement in APT [APT IV] defines pocket geometry, cutter geometry and feedrate.

A natural left (or right) preorder traversal of the manufacturing data tree may be used to define the machining sequence of the process plan. Thus, tree nodes may be used to generate modal commands. This reflects the syntax of block structured languages and provides the same advantages. However, the structure is inefficient when two (or more) nodes at the same level need to share a node at a lower level. This deficiency is analogous to that incurred by block structured programming languages when unnecessary repetition is occasionally required to preserve the elegance of the structure. For example, a change of setup need not necessarily involve a change of cutter. When drilling and tapping or rough and finish milling, the machined cavities are different but sufficiently closely related to be usefully defined as a single shared cavity. Thus, more than one cutter may be required to machine a single cavity of the manufacturing definition.

A process plan defined by the data described above is semantically machine tool independent. It is therefore portable between machines and suitable for archiving.   This contrasts with conventional part programs (including APT programms [APT IV]) and CLDATA files [EIA 1983], which are portable only to the extent that the target control systems require commands with a common syntax. In addition, such 'portable' programs may not have the desired effect; different machines may react differently to the same commands. The latest CLDATA standard [EIA 1983] is intended to improve portability by facilitating the design of machine tool/controller combinations which use the same CLDATA semantics; the system under discussion goes significantly further.   In theory, it allows manufacturing definitions to be ported to any machine tool which can produce the component. This results from the use of machining data which defines in chronological sequence what is to be achieved and the associated constraints, rather than how to achieve it.

Before commands are generated for a specific machine tool it is necessary for the manufacturing process definition to be checked for compatibility with a machine tool specification including feed rate and spindle speed ranges, geometric specification of the available machining region, and collision avoidance regions associated with the machine tool. Machine tool collision regions are those collision regions not modified by a change of setup. In the prototype implementation this includes the spindle structure and the table on which fixtures are mounted. Other collision regions are associated with the setup information. The specification of the machine tool controller command format is not relevant to generic machine command generation. It is relevant only to the generation of specific controller commands and in this context is discussed in section 5.

Machine tool data may be represented as an additional node at the top of the manufacturing data tree. This enables the transfer of a manufacturing process to a different machine tool simply by replacement of the machine tool node. Inclusion of the machine tool within the manufacturing data heirarchy also accommodates the manufacture of parts by operations on more than one machine tool. In addition, the structure is easily extended to accommodate a choice of machine tools in a system where flexible scheduling is required.

A formal (PASCAL) definition of manufacturing data as implemented in a volume modelling environment, and an example for the well known Gehause component is given in Middleditch & Stacey [1985]. This reference also contains an APT program to manufacture the same component using the process plan. Figure 4(d) illustrates the end product of the manufacturing process definition which was constrained by choice of cutters to make the component five times bigger than in the original design [CAM—I 1979].

3.0 DATA CHECKING

Various operations may be performed on complete and incomplete manufacturing data trees. These include checking for self consistency and the generation of graphical displays. The data checking module provides the facility to establish whether a manufacturing process definition can be executed without collisions or other violations such as feed rate, spindle speed or feed axis position out of range. This includes the ability of a specified cutter to remove a specific cavity, its ability to access that cavity, and the existence of a collision free path to an appropriate access point.

The data checking module updates the current workpiece with the specified cavities as the tree is traversed in order to check each stage of machining. It is therefore possible to produce a graphical display of the workpiece at any stage of the manufacturing process. This permits the user to verify that the manufacturing process definition will produce the desired component. Figure 4 illustrates some stages in the simulated production of the Gehause component.

Data checking and machine tool command generation require the interrogation of the data structure in a sequence consistent with the process plan, e.g. via a left pre-order tree traversal. The definition of the data structure implies that applications modules must be able to access all data on the path to the root of the tree from the node being processed.

Output from the data checking module is a set of error codes. In the prototype implementation these codes are simply interpreted and reported to the user.

3.1 Geometric Framework

The prototype implementation employs geometric methods first described in [Stacey & Middleditch 1986]. These methods use definitions of constraints on the motion of a 'cutter control point' within the machining region defined by the feed axis limits. The machining region, collision regions and other regions used to determine the path of the cutter, do not relate directly to the volume of space occupied by the cutter. They are the associated volumes occupied by the cutter control point in the machine tool configuration space, i.e. the space with coordinate axes corresponding to the machine's feed axes. In the case of Cartesian control the cutter control point is the fixed point on the cutter considered to be controlled by the NC system and the machining region is simply a volume in Euclidean 3-space.

Collision regions are the volumes occupied by the workpiece and objects fixed with respect to it (fixtures, etc.) offset by the volume occupied by the moving objects (cutter, cutter shank and cutter holder). Such offsets are computed using the vector difference operator [Middleditch & Morris 1984]. An offset cavity is the set difference between the cutting volume/initial workpiece collision region, and the cutting volume/required workpiece collision region. An access surface is the common boundary between the complement of the collision regions and the offset cavity.

The cutter control point must avoid collision regions to ensure that the moving parts of the machine

tool do not interfere with the static parts. It must also move on some sufficiently 'space filling' path within the offset cavity in order that the cutter interferes with and thus removes the required material. If the cutter control point enters an offset cavity via the access surface, the associated cutter enters the cavity without removing excess material or colliding with fixtures, etc. Similar undesired collisions during machining are detected prior to cutter path generation as interference between the offset cavity and a collision region.

The absence of undesired collisions together with the existence of an access surface guarantees the existence of a valid cutter path. A null offset cavity corresponds to a cutter too large for the required cavity, and a null access surface indicates that the offset cavity is completely enclosed within the workpiece collision region. Machining is not possible in either case.

The computation of exact constraints on cutter paths before cutter path generation yields significant processing time advantages over the generate and test method employed elsewhere (e.g. [Armstrong 1982]). Efficiency may be further enhanced if volume models representing the fixtures and initial workpiece, etc. are not exact. It is often sufficient to represent non-critical components by simple bounding volumes, e.g. cuboids.

3.2 Volume Models

The proposed geometric methods require the implementation of a vector difference operator to generate offsets of volumes with respect to other volumes. This operator has previously been implemented [Middleditch & Morris 1984] for convex two dimensional regions bounded by linear and circular arc segments. The extension to linear extrusions (lifts) of such areas is relatively simple. Therefore all volumes relevant to the machining process are restricted to such lift volumes. By convention, areas are assumed to lie in the xy-plane and are lifted up the z-axis. For efficiency and to retain the form familiar to the user, cylinders and cuboids are represented explicitly. It is also convenient to have a special representation for slot volumes as the vector sum of a cylinder and a piecewise path in the xy-plane. This facilitates the trivial evaluation of the vector difference of a slot and a cylinder (the cutter model).

Volumes are thus restricted to being two and a half dimensional (2.5D) and transformations associated with volumes (except viewing transformations) are correspondingly restricted. However, to permit future generalisation and compatibility with other software for display and mass property evaluation, the implementation employs more general data types for volumes and transformations than are necessary in the present context. Volumes are represented as a binary tree with set operators at the nodes and primitive volumes (including lift volumes, cuboids and cylinders) at the leaves. The transformations applied to these leaves are represented by 4x4 matrices.

It is further assumed that the moving parts of the machine tool (i.e. cutter, cutter shank and cutter holder) may be represented by cylinders. This enables vector sums of arbitrary areas to be replaced by vector sums of areas with discs (i.e. simple polygon offsets). It is then sufficient to support a suite of area manipulation utilities (including set operations and fixed radius offsets) analogous to those described in [Barton & Buchannan 1980].

3.3 Interference Checks

Having obtained the appropriate volume offsets, it is necessary to detect their interference with points, lines and other volumes. This may be achieved by the classification of a candidate point set against a second reference point set, to establish those points which are inside, outside and on the boundary. Since all volumes are lift volumes, these operations are separable into the equivalent operations on areas and lines.

Point classification may be performed by a test against the z–range of the lift volume and then, if necessary, the 2–dimensional classification of the projection of the point on the base plane. Similarly, line classification is performed by 2–dimensional classification of the projection on the base plane of that segment of the line lying in the z–range of the lift volume. Volume interference detection is performed by z–range overlap detection and the appropriate area interference detection. These projection methods for lift volumes are the subject of a further document [Middleditch & Stacey]. Area interference detection may be reduced to the interference of a point with the vector sum of areas [Middleditch & Morris 1984].

3.4 Regularisation

Volume modelling systems based on the set theoretic combination and transformation of primitive volumes usually employ a process known as regularisation [Requicha 1980]. This prohibits the formation of illegal volumes with, for example, hanging edges or faces. Regularisation is important primarily for design applications.

The system under discussion requires a vector difference offset operator. Using this operator a cylindrical volume may be offset inwards by another paraxial cylindrical volume to obtain an offset cylinder of positive or zero radius or a null volume. The degenerate case of zero radius could result from the offset of a hole by a cutter of the same radius; it should therefore not be prohibited by regularisation. A null offset, may be represented as a cylinder of negative radius if the offset is produced by naively subtracting radii, and corresponds to a hole which is too small for the cutter. Thus, the regularisation advocated for volume modelling systems may be undesirable in some applications.

Volume modelling systems should be able to represent non–regular volumes. Systems should either have both regularised and non–regularised construction algorithms, or alternatively, non–specific construction algorithms whose results are interpreted appropriately as regular or non–regular by the evaluation algorithms. In the former case, regularised construction algorithms prevent the formation of 'negative' volumes by equating them to null volumes as indicated by the vector sum definition for offsets. In the latter case, a zero radius cylinder may be interpreted as a null volume or a single line. A volume model may be considered to be defined procedurally by a point classification operator, and regularisation may be regarded as a parameter of the classification process. Offsets of volumes for collision detection and machining applications evidently require non regularised classification.

4.0 DATA COLLECTION AND ERROR RECOVERY

The data described above must ultimately be obtained from user input or generated automatically. In the former case users should have access to a predefined library of machine tools, standard cutters etc., which may be extended when necessary. In the latter case, the ultimate source of data should be solid models defined in a CAD system.

The prototype implementation obtains data relevant to a machining process from user input or from files named by the user. It is convenient to collect this data using a left pre–order tree traversal. Only manufacturing and descriptive geometric data need be entered since machine tool commands are generated internally. This eliminates the need for a user to submerge his machining experience in an intermediate programming language and should help to clarify those strategic decisions which can at present only be made on the basis of such experience.

Data in the input sequence is modal in the sense that data at a particular node remains relevant until the next node at that level is visited. This enables data to be checked on entry for compatibility with previously entered data, including the machine tool specification. In the event of an error the user may immediately modify the current data or any relevant previous data. Modified data is also checked for compatibility. Since correction of data can lead to further errors, error recovery is implemented recursively.

Errors may thus be corrected before any expensive machine tool operations occur or even before machine tool commands are generated. In particular, cutter collision, cutter/cavity compatibility and cutter/cavity access checks may be performed imediately the cavity and location are defined. This is made possible by the offsetting methods outlined previously. The usual response to an error detected at this stage would be to specify an alternative cutter or to change the setup, rather than to redefine the cavity.

5.0 GENERIC MACHINE TOOL COMMAND GENERATION

When the feasibility of machining has been established by data checking, there must exist a valid cutter path for the desired machining operation. Furthermore, geometric constraints on the cutter path are already available in the form of collision regions and offset cavities. While data checking, models of these regions are attached to the manufacturing data structure to avoid recalculation during cutter path generation. Hence data must always be checked before cutter path generation.

Machine tool commands are generated during a left pre-order traversal of the manufacturing data tree. Automatic work load commands (or positioning for manual loading and command suspension) are generated at set up nodes. Automatic tool change commands (or positioning for manual loading and command suspension) are generated at cutter nodes, and cutter paths are generated at location nodes.

5.1 Cutter Path Generation

An access path, avoiding collision regions, from the current position of the cutter control point to an accessible point of the offset cavity could be obtained using 3D path finding methods analogous to those described in [Lozano—Perez & Wesley 1979]. However, many NC machine tool controllers employ a rapid feed mechanism which is not readily susceptible to precise path control, so the present implementation uses a simple heuristic based on rapid feed at a safe height followed by cavity access parallel to the spindle axis of the machine tool. This is suitable for most simple applications.

No attempt is made in the prototype implementation to optimise cutter paths for material removal. Also, cavities are limited to a single linear extrusion, or lift volume. Thus, stepped cavities are modelled as the disjoint union of multiple lift cavities. This enables a material removal heuristic which ignores changes in cross section.

The material removal algorithm merely clears successive layers separated by a constant user prescribed depth of cut. Each layer (cross section of the offset cavity) is removed using cutter paths on the boundary of successsive offsets by a user prescribed width of cut. Thus, the cutter path spirals inwards. The direction of cutter motion may be clockwise or anticlockwise depending on whether the user specifies climb or conventional milling. If an offset operation results in disjoint regions, each region is successively completed.

The algorithm for machine tool command generation is illustrated in more detail by the pseudo PASCAL program in Middleditch & Stacey [1985].

5.2 NC Code Generation

Machine tool commands must be produced in the format expected by the machine tool controller. This is achieved by the NC driver module. In the current implementation the CAM system may be linked to a variety of machine tool driver modules with a common procedural interface [Morris & Middleditch 1984]. These modules produce machine tool commands in a range of formats, including the standard CLDATA format [EIA 1983]. The common procedural interface also permits linking to a graphical simulation module.

The role of the NC driver module is assumed to be analogous to that of a standard GKS graphics package; its function is to translate procedure calls into machine tool controller readable form. Its responsibility is for correct syntax of commands; the data checking module is responsible for their semantics. Since errors should be corrected before machine tool commands are produced, the NC driver should simply clip control point position, feed rate, etc. against their respective ranges in order not to produce illegal commands. However, if the NC driver is also to be used by less sophisticated machine tool command generation systems, it should be possible to inquire whether clipping has occurred. This provides a simple error checking facility. The procedural interface accommodates machine tools limited to two and a half axis control. Such machines have independent control parallel to the spindle (Z) axis and continuous path control orthogonal to the spindle axis in the XY plane. These capabilities are adequate for the manufacture of two and a half dimensional volumes. The NC command data generated for the component of figure 4 is recorded in Middleditch & Stacey [1985]. Figure 5 shows the cutter path output from the alternative graphical simulation driver.

6.0 SUMMARY

The CAM system described, together with a simple user interface, has been implemented on a VAX 11/780. A fairly large range of machined parts can be defined using the system and then manufactured on a Kearney and Trekker H60 machining centre. Although only orthogonal lift volumes are used for cavity definition, the workpiece transformation at setup level extends this domain beyond those objects usually termed 2.5D, provided that all cavities machined during a single setup are 2.5D.

The prototype system makes no attempt at the automation of process planning. Instead, a facility for the user to define his own process plan is provided. This approach permits greater flexibility and provides a basis for further research on the automation of various aspects of process planning. The latter is not entirely a geometric problem and requires significantly more information than can be obtained from purely geometric models.

The system differs in some significant respects from previous attempts at the automatic derivation of cutter paths using volume modelling methods (e.g. [Armstrong 1982]). Firstly, in the prototype implementation, cavities, cutters, obstacles, etc. must be 2.5D volumes. Secondly, offsetting methods enable data checking to be separated from machine tool command generation. This is more efficient than generating cutter paths, finding their swept volumes, then checking collision. It thus permits better response in an interactive system, and simplifies the derivation of cutter paths, since they are constrained to specific regions.

6.1 Acknowledgement

The example manufacturing process definition illustrated in figures 4 and 5 was created by Alistair Patterson.

7.0 REFERENCES

ARMSTRONG, G.T.: 'A Study of Automatic Generation of Non—invasive NC Machine Paths from Geometric Models', PhD Thesis, University of Leeds (1982).

BAILEY, C., MIDDLEDITCH, A.E.: 'Structured Information Flow for a Numerical Control Command Tape Preparation System', CAE Technical memo 78:3, PCL (1978).

BARTON, E.E., BUCHANAN, I.: 'The Polygon Package', Computer Aided Design (1980), Vol. 12, No. 1 P3—11.

APT IV: Encyclopaedia, IIT Research Institute, Chicago.

CAM—I, Proc. Geometric Modelling Seminar, Bournemouth, England, 1979, CAM—I pp 80—GM—01.

EIA Standard RS—494, Electronic Industries Association (1983).

LOZANO—PEREZ, T., WESLEY, M.A.: 'An Algorithm for Planning Collision—free Paths among Polyhedral Obstacles', Comm. ACM, (1979), Vol. 22, No. 10.

MIDDLEDITCH, A.E., MORRIS, A.C.: 'The Representation and Manipulation of Convex Polygons', CAE Technical Memo 84:6, PCL (1984).

MIDDLEDITCH, A.E., STACEY, T.W.: 'The Vector Sum in Set Theoretic Geometric Modelling', CAE Technical Memo 84:1, PCL (in preparation).

MORRIS, A.C., MIDDLEDITCH, A.E.: 'The Generation of Command Sequences for Numerically Controlled Machine Tools', CAE Technical Memo 84:7, PCL (1984).

REQUICHA, A.A.G.: 'Representations for Rigid Solids', Computing Surveys, (1980), Vol. 12, No.4.

STACEY, T.W., MIDDLEDITCH, A.E.: 'Implementation of a Computer Aided Manufacturing System', CAE Technical Memo 85:3, PCL (1985).

STACEY, T.W., MIDDLEDITCH, A.E.: 'The Geometry of Machining for Computer Aided Manufacture' Robotica, April 1986, Vol. 4, Part 2.

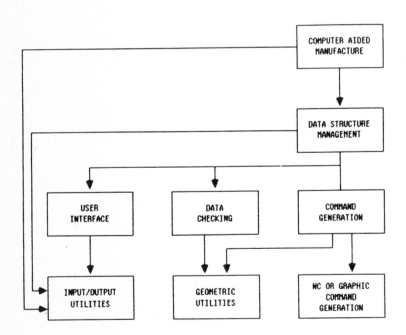

Fig 1   Software heirarchy

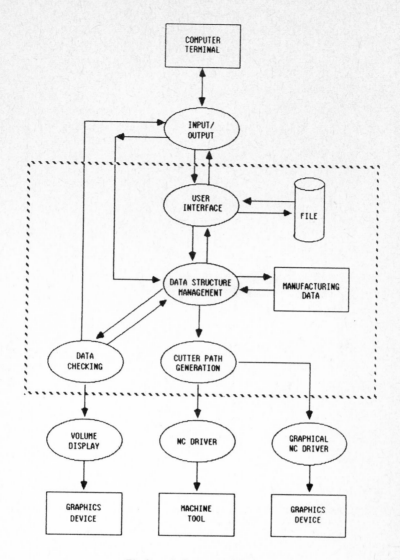

Fig 2   Information flow

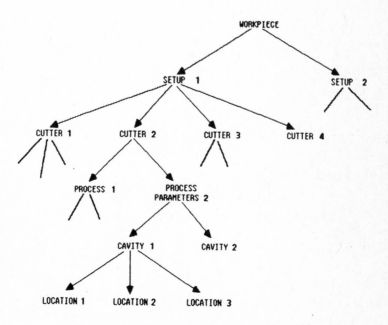

Fig 3   Manufacturing data structure

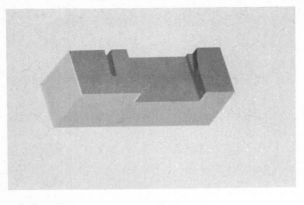

(a)

(b)

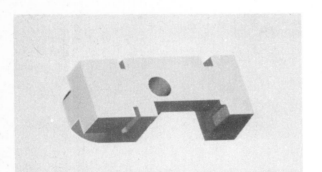

(c)

(d)

Fig 4   Four stages in the simulated production of the Gehause component
(a) End of the first set-up
(b) End of the third set-up
(c) Start of fourth set-up (rotated view of b)
(d) End of fourth set-up (finished component)

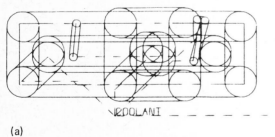

(a)

(b)

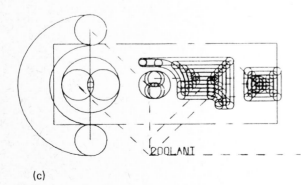

(c)

(d)

Fig 5   Graphical simulation of cutter paths for the Gehause component
(a) Set-up 1   (c) Set-up 3
(b) Set-up 2   (d) Set-up 4

C362/86

Engineering trade-offs in the design of a real-time system for the visual inspection of small products

E R DAVIES, MA, DPhil, CPhys, MInstP
Department of Physics, Royal Holloway and Bedford New College, Egham, Surrey
A I C JOHNSTONE, BSc
Department of Computer Science, Royal Holloway and Bedford New College, Egham, Surrey

SYNOPSIS This paper presents an algorithm for the rapid inspection of small products, and considers its optimal implementation. Speed and cost limits constrain optimisation relatively simply, but concurrency makes the situation considerably more complex. However, an inspection system that makes particularly efficient use of a set of hardware processors has been designed. The host CPU is not merely used for control and data logging, but takes an integral role in the main image analysis task. The emphasis of the paper is on removing arbitrariness in the design of hardware for industrial inspection systems.

1 INTRODUCTION

The past decade has seen enormous growth in the applications of computers to manufacturing engineering. Automated assembly and automated inspection are becoming mature technologies: both customarily use vision, since sensors such as TV and linescan cameras are capable of providing prodigious quantities of relevant information at high data rates, and algorithms are known which will analyse much of this data sufficiently reliably for the purpose of control. In automated assembly, vision provides data on the positions and orientations of products, and can at the same time check them for defects. In automated inspection, the immediate purpose is that of dimensional checking and quality control. More specifically, it aims (a) to identify and reject products containing defects, (b) to provide feedback on product characteristics (such as size or surface texture) so as to keep manufacture within specified tolerances, and (c) to provide statistics on manufacturing parameters (1).

There are two important trends in manufacturing: one is that products are becoming increasingly complex and sophist-icated; the other is that the consumer is becoming significantly more demanding with regard to quality. As a result there is a move towards 100 per cent inspection of products.

If there is to be 100 per cent control of quality at a time when products are becoming increasingly complex, this necessarily throws a heavy load on computing machinery. Indeed, the enormous amounts of data in typical images, coupled with the high throughput rates, mean that special electronic hardware is needed for visual inspection. Such hardware is costly, and there are two basic strategies for providing it: the first is to maintain generality by building multi-processor systems containing (for example) large numbers of microprocessors; the other is to employ special dedicated hardware systems which are less adaptable but may be considerably cheaper in individual applications. This paper studies these basic strategies and analyses the speed/cost tradeoffs in hardware for automated visual inspection. In addition, it describes the real-time system we have designed to inspect products such as biscuits at costs compatible with the low profit margins of foodproduct manufacture.

2 ALGORITHMS FOR PRODUCT INSPECTION

Many inspection problems involve three main tasks: (1) image acquisition, (2) product location, and (3) product scrutiny and measurement. In this section we bypass the problem of acquisition and concentrate on product location and scrutiny.

To locate products in a grey-scale image it is very common to threshold the intensity and thus obtain a binary image from which objects can be located with relatively little further processing. This scheme is only suitable if the lighting system is carefully configured and products appear silhouetted e.g. as dark objects against a light back-ground, so that the intensity histogram is bimodal. Many inspection systems have been designed on this basis, but attention is now shifting towards more complex products for which this approach is unlikely to be successful. For this reason we concentrate here on systems based on edge detection. This approach is generally much more robust, being able to negotiate problems due to shadows, overlapping products, etc.

One common approach to object location is to use a simple edge detector to locate the boundaries of objects, and then to link broken edges in order to create complete (connected) outlines of objects. Having obtained complete object outlines, these can be thinned down to single pixel width, and a tracking algorithm can be used to follow the outlines of individual objects, hence generating a set of polar (r,θ) plots (1,2). Efficient one dimensional pattern matching algorithms can then be used to identify objects, at the same time determining their orientation by finding how much the observed profile needs to be shifted before it matches a standard template. It is worth noting that object scrutiny can then proceed, at least in part, by checking how close the match with the standard template is.

This approach, which we shall call Algorithm A, has the disadvantage that image noise, and the nature of the objects and the lighting, together with possibilities of overlapping objects and other artefacts, can make it difficult to link broken edges together. Indeed, there are arguments that indicate that it is not in general possible to perform this function successfully: in any case we have found many practical instances of failure. However, in the restricted set of cases in which the algorithm can be used, it operates extremely rapidly, since the tracking algorithm need not visit every pixel but can by-pass much of the image. An unfortunate consequence of this is that the algorithm does not take in enough data to be particularly robust. In our work, we have aimed at high accuracy and extremely high levels of robustness, since these are the qualities we have found particularly in demand in industrial applications of vision: we have therefore avoided the tracking schemes used by Algorithm A.

An alternative approach which we shall call Algorithm B also involves edge detection, but instead of linking broken edges together directly it proceeds with the Hough transform strategy of locating objects by accumulating (at a reference point within each object) the evidence that is available for their existence (3,4). This gives a particularly robust strategy for locating objects, though it is not so fast as Algorithm A. However, Algorithm B is still highly efficient, since full use is made of locally available edge orientation information to compute candidate positions for object reference points. Here we illustrate this technique by reference to circular object location. For a circular object, each edge pixel that is found permits computation of a candidate centre point a distance equal to R (the radius) along the edge normal in 'parameter space' (4). (In this application parameter space is isomorphic to image space.) When all edge pixels in an image have been processed, it is only necessary to search parameter space for clusters of candidate centre locations and to average them to find accurate positions for the centres of circular objects.

Once products have been found using Algorithm B, they may conveniently be scrutinised using the radial histogram method, which involves computing an average radial intensity profile of the product, and matching this against a suitable template (5,6). This is again a one-dimensional approach, and is analogous to that of matching one-dimensional shape profiles in Algorithm A, though it aims to provide information on product reflectivity and size rather than shape (5-7). Naturally, Algorithm A could also be augmented by the use of radial intensity histograms, if this were appropriate.

These arguments show that Algorithm B should be considerably more robust and accurate for object location than Algorithm A. We have made extensive practical tests of the situation, particularly for the location of round foodproducts such as biscuits, and have verified that this is so. Specifically, Algorithm B has been found to be exceptionally tolerant of broken and overlapping products, and those having other shape defects such as protuberances around their edges (5). In what follows we assume that Algorithm B is to be used because of its superior robustness and accuracy.

3 THE SPEED PROBLEM

For our tests on biscuit inspection, Algorithm B was augmented to include assessment under four main headings - roundness, radius, amount of chocolate cover, and general acceptability according to a radial intensity correlation coefficient. The initial version of this algorithm took about one minute to run on a PDP-11/34A. Subsequent optimisation of the algorithm strategy brought the execution time down to ~5 seconds, without resorting to hand massage of machine code. Although the original algorithm was written in Pascal, it became clear that attempts to optimise the machine code would (for this algorithm) result in a speedup factor of less than two: a 3 second overall execution time appeared to be a limiting case. This is to be expected for the following reasons. The edge detection part of Algorithm B requires some 16 image accesses for each pixel in the 128x128 image. Although the access time of the framestore is around 1 microsecond, instruction fetches and overheads within the program loops reduce the average throughput to around 100000 pixels/second. As a result the minimum execution time of the edge detector is of the order of 2.6 seconds. Product scrutiny then requires some 4-5 accesses over the relevant area (about 3000 pixels in our application), which will take at least another 0.1 seconds. The Hough transform calculations are only applied to some 200 points, but their high computational cost will add another 0.1 seconds to the total execution time. Clearly even with ideal code generation there is a lower bound on the overall processing time of about 2.8 seconds. Changing to a 68000 or other commonly available microprocessor would not affect this substantially.

At best, software optimisation is subject to severely diminishing returns, and further speedup must rely on enhancement of the hardware implementation. As stated in section 1, this has to be obtained either by use of several central processor units (CPUs) or by specially designed dedicated electronic hardware. To inspect biscuits at typical rates of 10-20 per second, a speedup factor ~100 must be attained.

For industrial applications, cost has to be kept low, and it is useful to see how generality can be maintained subject to this

16

constraint. With this in mind we examine a number of alternative processing architectures.

4 MULTIPROCESSOR SYSTEM DESIGN

4.1 SIMD architecture

When considering fast hardware for image analysis applications, it is natural to start with the SIMD machine, since this architecture would appear to match the hardware to the algorithm most accurately (8). The SIMD (or 'Single Instruction Multiple Data') architecture when applied to image analysis ideally involves use of one processing element (PE) per pixel, the PEs being arranged in an array isomorphic with the image being processed. Such a machine is able (for example) to invert or threshold an image in one instruction cycle, since all processors operate simultaneously on their respective pixels. This type of machine is also able to operate rapidly on images to remove noise by local averaging, or to find edge pixels rapidly by operations within 3x3 neighourhoods. In addition, it can efficiently build up distance functions or find object skeletons by sequences of 3x3 neighbourhood operations. A typical SIMD machine (9) is able to perform these 3x3 operations efficiently since each processing cell has direct links with its 8 neighbours (only 4 in the case of some machines such as the ICL DAP (10)), so the required data is immediately available.

Most SIMD arrays include an activity bit for each PE which allows selective application of processing steps to different areas of the image by disabling individual PEs. However, this clearly wastes the power of the SIMD machine. Unfortunately, all but the lowest level image processing operations require selective processing of the image. The Hough transform calculations required for Algorithm B form an interesting extreme case in which around 200 special image points (edge pixels) trigger a floating point calculation, at the end of which a single point in parameter space must be accessed. In principle the floating point calculation could be performed at every point by the SIMD array, but the only available way of performing the subsequent random access of the parameter plane is by propagation techniques (9). Typically 40 propagation cycles per point would be required in this application for each of the 16384 pixels in the image. (Note that an attempt to re-organise Algorithm B so that propagation routines are used to locate circle centres leads to significant loss of generality, since Algorithm B itself is immediately generalisable to detect any object shape (4).) A conventional sequential processor would be slow at calculating the edge image, but could then efficiently execute the 200 floating point calculations and directly access the parameter space. In addition, technology constraints dictate the use of simple bit-serial processors in current SIMD machines, and these would require many cycles to execute the required floating point calculation. Clearly, the pure SIMD solution would be much less efficient.

An alternative hybrid strategy would be to perform the edge calculation in the SIMD array, and then to read the results out sequentially into a conventional processor which would perform the floating point calculation and update the parameter space. The economics of this approach would be dictated by the relative costs of the SIMD and sequential machines and the bandwidth of the communication channel. Current SIMD arrays are still rather expensive devices, which discounted their use in our application.

This analysis shows that SIMD architectures are of limited use for processing tasks that cannot efficiently exploit their regular topology. The simplicity of the individual PEs, and the absence of long distance communication links within the image make them particularly unsuitable for geometrical calculations on object features. Thus the SIMD architecture is currently inappropriate for many tasks of image analysis that might be needed in industrial inspection, even though it might be well adapted to various image processing tasks in a general imaging environment.

4.2 Multi-processor systems

General multi-processor structures provide resources that may be used concurrently in an unrestricted fashion, unlike the SIMD machine where all resources operate in lockstep. As with all forms of parallel implementation, the efficiency of a multi-processor system will be dictated by the effectiveness of the functional partitions. Interactions between functions will require either transmission of data between processes or access to shared memory spaces. In the one case there is a potential data bottleneck due to lack of bandwidth in the communications channel, and in the other, processes may stall during contention for shared memory. Therefore the speed of a multi-processor system containing N processors is never increased by the ideal factor N unless there is no process interaction, which is unlikely to be the case in a system doing useful work. High efficiency will be obtained by minimising process interaction. Naturally, there is the risk that a system containing N processors and capable of increasing speed by the factor ~100 noted in section 3 will be rather an expensive solution.

4.3 Pipelined processing systems

Pipelined processing systems form an interesting sub-class of multi-processor systems which can be useful for the repetitive execution of a given set of operations. This is typically the case for industrial inspection systems, where the same algorithm is applied to each frame of data as it comes off the camera. In a pipeline, individual frames of data are passed along a chain of processors so that in an N-processor system, N different data sets are being processed at any one time.

Since all processors pass their completed data set on up the chain at the end of a fixed time slot, pipelines are only as fast as the slowest processor in the chain. To be optimal, all procesors should complete in the same amount of time. For a video-based system, an obvious approach would be to execute in integral numbers of TV frames. For high level parts of the algorithm, such as the Hough transform calculations, subdivision into equal execution time processes would be virtually impossible to achieve. Finally, the approach requires significant bus switching logic and local memory, as well as the hardware processors, which are themselves liable to be costly. Thus pipelined systems pose a serious partitioning problem, and in addition to lacking generality are likely to constitute a rather expensive solution to the speed problem.

4.4 General processing capability

We have concluded in our work that, contrary to many of the suppositions about image analysis (based on what is frequently valid in image processing per se), the ideal type of processing system is a highly general multi-computer system, which is abstract in the sense of not being tied to any specific imaging representation. Again this is not achievable within the budget of most industrial inspection systems. For algorithms such as Algorithm B, the best compromise seemed to be to make optimum use of a single CPU by linking it with a set of hardware accelerators selected for maximal generality coupled with applicability to the problem in hand. In this context, Algorithm B was seen as constituting a useful case study in algorithm analysis and multi-processor system design: this will be discussed in more detail below.

5 FURTHER ANALYSIS OF ALGORITHM B AND ITS IMPLEMENTATION

Table 1 gives a breakdown of the functions in Algorithm B. The 'description' indicates the size of the neighbourhood employed in imaging operations. It also indicates those processes that are one-dimensional: these are marked since they involve loops containing a significant number of operations, but not as many as for two-dimensional image processing in 1x1 or 3x3 neighbourhoods.

The two other headings in the table, time for execution in software on an LSI-11/23 and cost of hardware implementation, are somewhat notional since it is difficult to divide the algorithm rigorously into completely segregated sections. For example, it has been assumed that various overhead costs such as that of a backplane, rack and power supply have already been covered: we shall largely ignore such complications in what follows. Overall, the figures presented here should be sufficiently accurate to form the basis for useful decisions on cost effectiveness of hardware. Finally, costs are based on chip and other component prices, and do not include logic design or p.c.b. layout. However, on the whole the cost of design and layout work is proportional to the number of connections, which is itself roughly proportional to component cost. This means that our results will be substantially correct, since the analysis below is independent of scaling.

As a simple starting approximation, any function that is implemented in fast hardware will be assumed to run in zero time. To find the most cost-effective means of speeding up the system, we should therefore consider a sequence of options in each of which one additional function is implemented in hardware, successively reducing the load on the host CPU. To achieve this systematically, we should examine the speed-cost product (or cost/time ratio) of every function, and in successive options implement in hardware the function currently having the lowest value of this parameter: the rationale for this is to preferentially replace in hardware those functions that are slow and whose cost is relatively small, by applying a criterion function with suitable weighting values.

This simple procedure is made somewhat more complex by the significant economies that are possible when implementing functions 6-10, e.g. by using common pixel scanning circuitry. Specifically, any subset of the functions 6-10 can operate with a single interface, scanning circuit and radial position lookup table (which gives a value for radial position once x and y displacements relative to the circle centre are known). On the other hand, any subset of these functions that is not implemented in hardware engenders a time overhead in software. A full analysis of the problem would require a large number of functional partitions to be examined in order to find the optimum system configuration. However, this exhaustive search procedure need not be performed in this instance since the time overhead is much greater than the sum of the software times for functions 6-10. This means that once the initial cost overhead has been paid it will clearly be optimal to implement all of these functions in hardware. For this reason we group functions 6-10 together in the remainder of this paper. Table 2 summarises the position.

Table 2 shows that the cost/time ratios divide themselves into four main categories: (1) those of the order of 1 £/ms which are clearly worth implementing in hardware; (2) those between ~5 £/ms and 25 £/ms which will also have to be implemented in hardware to get a reasonable speed system; (3) those around 100 £/ms which it would be worth implementing if a very much faster system were needed; and (4) those above 1000 £/ms which it would probably never be economical to implement in hardware. If option 1 were chosen, the total cost of the system would be £9000 and the algorithm would run in 0.7 seconds; if option 2 were used, the system would cost £13700 and would run in 0.1 seconds; if option 3 were chosen, the system would cost £23700 and would run in 0.002 seconds, whereas with option 4 the system would cost £27700 and would run in zero time (in the current approximation). Here we have assumed that the base cost of computer plus camera, frame store, backplane, power supply, etc is some £6000 and that this will permit the algorithm to run in ~5.0 seconds as indicated in Table 2.

In the above analysis we assumed that those functions implemented in electronic hardware run in zero time. This will not be entirely valid in practice, and the most serious errors will be for image neighbourhood operations - particularly those for neigh-

18

bourhoods of size 3x3. Taking 150 nsec as the fastest time for pixel access (as with our implementation using the VME bus), we see that a 3x3 neighbourhood operation in a 128x128 image takes some 25 msec. With suitable local storage this could be reduced to ~8 msec or even to ~3 msec. For a 1x1 neighbourhood, pixel access times would be ~3 msec. Next, let us assume that the actual processing is carried out by TTL circuitry in some tens of nano-seconds per pixel location; then the processing time will be less than 1 msec. Thus quite straightforward circuitry could be used to implement each function in times as short as 3-4 msec: this goes some way to justifying, and extending, the approximation we made earlier.

We now interpret our finding that the cost/time ratios fall into four main categories. Broadly, the first category (cost/time ~1 £/ms) arises for imaging operations in 3x3 neighbourhoods, which are well worth implementing in hardware. The second category (cost/time in the range 5 to 25 £/ms) arises for faster imaging operations in 1x1 neighbourhoods. The third category arises for one-dimensional operations which involve less processing, and certain rather time-consuming floating point operations. And the fourth category is a general data processing category with non-repetitive operations that run so fast they are unlikely to be worth implementing in dedicated hardware. Specifically, functions 5,12,13,14 require tedious logic and/or floating point arithmetic, which means that one is competing with the cost-effectiveness of mass-produced CPUs if one implements them in hardware: in general it is not worth doing this.

Function 4 is at the high end of category 2 since it involves relatively few pixels and is essentially a one-dimensional rather than an imaging operation: in addition, its cost is rather high because it performs quite complex arithmetic.

5.1 More rigorous investigation of hardware-software tradeoffs

We now attempt a more rigorous analysis of the effectiveness of implementing the various functions in hardware. A complete breakdown of the overall cost/time ratio sequence is given in Table 3. t and c are the times and costs of the functions. Assuming an overhead cost of £6000 (see above), T and C are the overall times and costs resulting from implementing in hardware all functions down to the one indicated: the minimum value of T is taken as 0.030 seconds and is based on realistic values for the imaging and 1-D operations, as discussed earlier. Looking at the C*T product should now give an indication of the optimal tradeoff between hardware and software: this occurs for 13 functions implemented in hardware.

It is important to realise that minimising the C*T product only gives a general indication of the required hardware-software tradeoff. A lot depends on the original specification for the inspection system: it might be that the main aim is to meet a certain cost or speed

rather than to produce a 'bargain package' that might do well in the market place. In our work, we have aimed particularly at foodproduct inspection, where it seemed to be vital to minimise costs while keeping speeds moderately high (5). For this reason, we aimed at an overall cost of less than £10000. By implementing functions 1,3,6-11 in hardware, we found we could get within a factor 3.6 of the optimal tradeoff (C*T product). However, another important factor arose in this analysis: that was the declining cost of faster CPUs. Table 4 shows the same C*T calculation for an LSI-11/73 host processor replacing an LSI-11/23. In this case the optimum tradeoff again occurs for 13 functions implemented in hardware. However, our compromise of implementing only functions 1,3,6-11 in hardware is now within a factor 1.8 of the optimal tradeoff. It seems fair to assume that these factors will become even more attractive with future CPUs.

5.2 Further factors in hardware design

Some further improvement in performance was obtained by making use of the fact that the host processor and the dedicated hardware can operate concurrently. (Ideally we would gain a factor two in speed by using two processors, but it is clear that our design criteria involve mandatory partitions in the algorithm which are inimical to such a large gain in speed.) In particular, we found that function 2 can run in the host CPU while function 3 runs in hardware, and function 13 can run on the CPU while functions 6-11 run concurrently in hardware. Figure 1 gives an execution map of our implementation, showing that our final allocation of functionality to hardware and software is able to make significant gains in efficiency and speed. This further justifies implementing relatively few functions in hardware.

In our implementation of Algorithm B, we have achieved 25 msec for function 3 (edge detection), and 10 msec for functions 6-11: we are currently upgrading these to roughly double the speeds. At that stage the timings will be as indicated in Figure 1, and at a total cost of £12500 (using an LSI-11/73 with functions 1,3,6-11 in hardware) we will have a system capable of inspecting 11-12 products/second using Algorithm B.

5.3 Generality of the functions implemented in hardware

Algorithm B was partitioned into sections that correspond to a significant degree of generality. First, edge detection itself is a highly general image analysis function (11); second, the Hough transform procedure used for object location is generalisable to a variety of shapes (4); third, the radial histogram approach has the potential for being used even in cases where cylindrical symmetry does not exist, since it can be used to provide a rotationally invariant 'signature' character-istic of one or other part of an object in the region of an easily locatable feature. Finally, certain thresholding operations (e.g. counting the number of pixels darker or lighter than certain threshold values) are exceptionally easy to implement yet generally useful for object scrutiny.

Clearly, function generality is a crucial factor which will frequently override the C*T criterion in deciding on the priorities for building dedicated hardware. We have kept this in mind while deciding which functions to implement in hardware in our visual inspection work.

6 CONCLUSION

This paper has presented an algorithm for the rapid inspection of small products such as biscuits. It has analysed how this algorithm may optimally be partitioned between dedicated hardware and software. Detailed specifications such as strict speed or cost limits have been seen to constrain the basic optimisation procedure, and function generality is also a critical factor. In addition, it has been found difficult to decide systematically the best ways of incorporating concurrency into the design when processors take radically different forms: however, we have been able to design an inspection system that makes efficient use both of the host CPU and of a limited number of hardware processors. The approach we have adopted seems somewhat unusual in that we have proved it best to retain use of the host CPU for a proportion of the processing rather than to set about building everything in dedicated hardware: specifically, the host CPU is not merely used for control and general data logging, but is used to take an integral role in the main image analysis task. Ultimately, the aim of our work is to develop the methodology of digital hardware design for industrial inspection applications, and at the same time to arrive at optimal designs rather than ones that contain arbitrary sets of ad hoc processors.

Acknowledgements

The authors are grateful to the SERC and to United Biscuits and Unilever for financial support during the course of this work.

REFERENCES

(1) DAVIES, E. R. A glance at image analysis - how the robot sees. Chartered Mechanical Engineer, Dec 1984, 32-35

(2) PARKS, J. R. Industrial sensory devices. Ch. 10 in BATCHELOR, B. G. (editor) Pattern Recognition - Ideas in Practice. Plenum: New York, 1978, 253-286

(3) HOUGH, P. V. C. Method and means for recognising complex patterns. US Patent 3069654, 1962

(4) BALLARD, D. H. Generalising the Hough transform to detect arbitrary shapes. Pattern Recognition, 1981, 13, no.2, 111-122

(5) DAVIES, E. R. Design of cost-effective systems for the inspection of certain foodproducts during manufacture. Proc 4th Conference on Robot Vision and Sensory Controls, London, 9-11 Oct 1984, 437-446

(6) DAVIES, E. R. Radial histograms as an aid in the inspection of circular objects. IEE Proceedings D, 1985, 132, no. 4, Special Issue on Robotics, 158-163

(7) DAVIES, E. R. Precise measurement of radial dimensions in automatic visual inspection and quality control - a new approach. in BILLINGSLEY, J. (editor) Robots and Automated Manufacture, IEE Control Engineering Series 28. Peter Peregrinus Ltd: London, 1985, 157-171

(8) DAVIES, E. R. Image processing - its milieu, its nature and constraints on the design of special architectures for its implementation. in DUFF, M. J. B. (editor) Computing Structures for Image Processing. Academic Press: London, 1983, 57-76

(9) FOUNTAIN, T. J. CLIP4: a progress report. in DUFF, M. J. B. and LEVIALDI, S. (editors) Languages and Architectures for Image Processing. Academic Press, London, 1981, 283-291

(10) HUNT, D. J. The ICL DAP and its application to image processing. in DUFF, M. J. B. and LEVIALDI, S. (ibid) 1981, 275-282

(11) DAVIES, E. R. Circularity - a new principle underlying the design of accurate edge orientation operators. Image and Vision Computing, 1984, 2, no. 3, 134-142

Table 1 Breakdown of algorithm B

	function	description	time	cost	c/t ratio
			(sec)	(£)	(£/ms)
1.	acquire image	1x1	–	1000	–
2.	clear parameter space	1x1	0.017	200	11.8
3.	find edge points	3x3	4.265	3000	0.7
4.	accumulate points in parameter space	1x1	0.086	2000	23.3
5.	find averaged centre	–	0.020	2000	100.0
6.	find area of product	1x1	0.011	100	9.1
7.	find light area (no chocolate cover)	1x1	0.019	200	10.5
8.	find dark area (slant on product)	1x1	0.021	200	9.5
9.	compute radial intensity histogram	1x1	0.007	400	57.1
10.	compute radial histogram correlation	1-D	0.013	400	30.8
11.	overheads for functions 6-10	–	0.415	1200	2.9
12.	calculate product radius	1-D	0.047	4000	85.1
13.	track parameters and log	–	0.037	4000	108.1
14.	decide if rejection is warranted	–	0.002	4000	2000.0
	time for whole algorithm		4.960		

Table 2 Revised breakdown of algorithm B

	function	description	time	cost	c/t ratio
			(sec)	(£)	(£/ms)
1.	acquire image	1x1	–	1000	–
2.	clear parameter space	1x1	0.017	200	11.8
3.	find edge points	3x3	4.265	3000	0.7
4.	accumulate points in parameter space	1x1	0.086	2000	23.3
5.	find averaged centre	–	0.020	2000	100.0
6-11.	set of functions with same overhead	1x1	0.486	2500	5.1
12.	calculate product radius	1-D	0.047	4000	85.1
13.	track parameters and log	–	0.037	4000	108.1
14.	decide if rejection is warranted	–	0.002	4000	2000.0
	time for whole algorithm		4.960		

Table 3 Speed—cost trade-off figures for LSI-11/23 based system

order	function	t	c	T	C	C*T
		(sec)	(£)	(sec)	(£)	(£-sec)
0	–	–	6000	4.990	6000	29940
1	3	4.265	3000	0.725	9000	6530
2	6-11	0.486	2500	0.239	11500	2750
3	2	0.017	200	0.222	11700	2600
4	4	0.086	2000	0.136	13700	1860
5	12	0.047	4000	0.089	17700	1580
6	5	0.020	2000	0.069	19700	1360
7	13	0.037	4000	0.032	23700	760
8	14	0.002	4000	0.030	27700	830

Table 4 Speed—cost trade-off figures for LSI-11/73 based system

order	function	t	c	T	C	C*T	gain
		(sec)	(£)	(sec)	(£)	(£-sec)	
0	–	–	7000	2.154	7000	15080	1.99
1	3	1.835	3000	0.319	10000	3190	2.05
2	6-11	0.207	2500	0.112	12500	1400	1.96
3	2	0.006	200	0.106	12700	1350	1.93
4	4	0.035	2000	0.071	14700	1040	1.79
5	12	0.017	4000	0.054	18700	1010	1.56
6	13	0.016	4000	0.038	22700	860	1.58
7	5	0.007	2000	0.031	24700	770	0.98
8	14	0.001	4000	0.030	28700	860	0.97

The last column in this table shows the overall gain in speed relative to the corresponding LSI-11/23 option in Table 3.

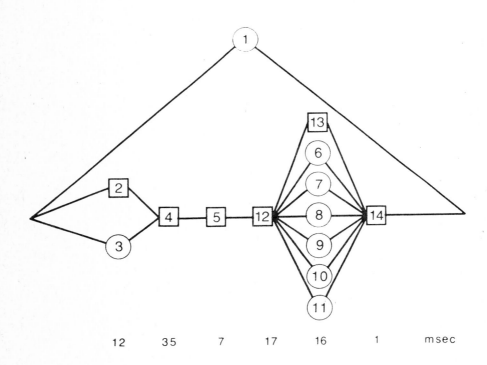

| | 12 | 35 | 7 | 17 | 16 | 1 | msec |

Fig 1   Execution map of algorithm B showing its implementation, making use of:

(a) pipelining of image acquisition and algorithm execution;
(b) simultaneous execution in hardware and software;
(c) sharing of scanning overhead and data I/O for functions six to ten.
Also indicated are the execution times of individual processes totalling 88 ms

☐ operations involving host CPU
○ operations executed in dedicated hardware

C370/86

A fast yet very low cost vision system

P J O'NEILL, BSc, MIERE
IBM (UK) Laboratories Limited, Winchester, Hampshire

SYNOPSIS A cost-effective application of machine vision to the 100% inspection of small piece parts
is described. The problem and the solution demonstrated represents, in the author's opinion, one of
the few applications of machine vision where no alternative method could have achieved satisfactory
results.   The advantages of customising the vision hardware and algorithms and of providing a high
level of software are shown.

1. THE PROBLEM

A common source of praise from IBM's customers
is the quality and 'feel' of their keyboards.
The component that is probably most responsible
for this sense of quality is the spring pad
assembly. This part (Figure 1) is a small,
high precision, plastic moulding with a
stainless steel coil spring force-fitted onto a
peg. One of these assemblies is fitted under
each key button – in some models more than one
hundred per keyboard. The coil spring fits into
a recess under the keybutton; depressing the
key compresses the coil spring but the geometry
is so arranged that the forces are not co-axial
with the spring and at a certain compression
the spring buckles. The buckling exerts a
torque on the plastic pad which then rotates
from its normal rest state to a tilted position
– this gives the familiar 'click' and the
satisfying positive feel to the keyboard
(Figure 2) and, of course, operates the
electrical circuit.

Unfortunately, to produce consistent
behaviour, in what is a complex non-linear
mechanical system, requires tight control of
many of the part's characteristics. Most
tolerances can be addressed by good control of
the pad moulding and of the spring winding
using normal inspection and sampling methods.
However, when the springs are pressed onto the
pads in the assembly machine, a number of
parameters need to be controlled that are both
difficult to monitor on-line and not amenable
to sampling techniques due to the known,
somewhat inconsistent, behaviour of the
assembly machine. The two important aspects to
be controlled are the height of the assembly
from the base of the pad to the top of the
spring and the orientation of the cut end at
the top of the spring. Height is closely
specified since it directly affects the degree
of key depression at which the buckling takes
place, the orientation has to be well
controlled so that the cut end will fit into a
relieved area in the key button – if the
bearing point of the key is on the cut portion
of the spring it can lead to erratic operation,
key button wear and production of debris.
Figure 3 shows the two tests.

Given the vast quantities of these
assemblies that are produced and that the
assembly machines are liable to random errors
in operation, a fully automated, high-speed,
100% inspection system is clearly essential.

2. PREVIOUS APPROACHES

Many millions of spring pads have been made and
fitted into keyboards without any complete
inspection system. Height and orientation have
been checked by classical batch sampling but it
was known that there were escapes and with
upwards of a hundred parts in each keyboard a
very low escape rate could have a quite large
impact on overall product quality levels,
causing very expensive rework of fully
assembled keyboards during their final tests.
Furthermore, sample tests that did reveal
out-of- specification parts resulted in
quarantining of large batches of parts, most of
which were good but not economically or
practically separable from the small number of
genuinely bad ones. It was the ever increasing
inventory of suspect parts that provided the
final impetus towards finding a solution that
would give complete inspection.

A cursory examination of the parts shows
that any mechanical in-contact tests would be
very difficult to engineer reliably – the small
size and the non-rigid nature of the spring
ensure this. The assembly machines are already
fitted with one mechanical test, a simulated
key depression station. At this station the
completed assembly is depressed several times
by a cam-operated plunger and the 'clicking' of
the pad is detected by a proximity sensor;
this would appear at first sight to be a
complete functional test but, unfortunately, it
does not detect bad orientation effectively and
it will not discriminate well enough parts that
are outside of the height specification. This
'buckle test' does, however, catch some parts
that, whilst appearing to be in specification,
still do not deform reliably. Adding height
and orientation inspection to the machines
should therefore be complementary to the
functional test.

Rejecting the possibility of a mechanical inspection method leaves only some form of electronic sensing or an optical system. The height could probably be readily checked by a capacitance probe but it is difficult to see any easy electronic way of sensing the orientation. However, since the two measurements are both morphological aspects of the top of the spring then a visual method immediately suggests itself.

3. EARLY WORK

The UK Development/Manufacturing Process Centre (UK DMPC) at IBM's Hursley Laboratory were initially asked to look at two related problems: how to extract good parts from the ever increasing stockpile of suspect batches that was building up during 1983/84 and how to improve the orientation sampling process.

Preliminary static tests showed that it was possible to obtain good silhouette images of the tops of the springs and that a simple geometrical algorithm could determine the angle of the spring end. From these tests we went ahead to produce a single special purpose piece of test equipment that was attached to an existing spring pad tester (with the buckle test station mentioned above).

This equipment used a fairly low resolution solid-state array camera (128 x 128 pixels), a commercial camera control and light strobing box, a small purpose built logic and interface card and was attached to an already available IBM Series 1 computer. Only the orientation test was performed.

Most of the development work was concentrated into solving the problems of getting reliable images, in devising the optimal hardware and software algorithms for reducing and analysing the image data and in integrating the electronics into the mechanics of the testing machine. This tester was brought into use at IBM's Greenock plant in 1984; it was used to re-test some millions of suspect parts from which 90 percent good parts were recovered. It has also subsequently been in regular use as the orientation sample tester.

Results were extremely encouraging – the salvaging of millions of good parts from the 'bad' stock, with high reliability, led at once to the request to develop a production version able to be fitted to every assembly machine and able to perform both tests on line.

4. CONCEPTS

As mentioned above, the original re-test machine was produced somewhat as a 'fire-fighting' exercise to diminish an expensive stock of parts that had had value added yet at the same time had no realisable value. The plan to fit a vision 'inspector' to each assembly machine required a quite different approach to engineering and implementation. UKDMPC and the Greenock plant agreed some design objectives that are summarised below.

a. The test station had to be easily added to the existing assembly machine and should not substantially increase the machine's physical envelope. The functional test station ('buckle tester') should be retained.

b. The machine should function as a stand-alone tester yet be able to talk to a host system for collection of statistical data. A wide range of statistical measurements had to be kept.

c. Both orientation and height should be tested.

d. Low cost was essential (the original target was £6000 per unit).

e. Operational control of the assembly machine was to be transferred to the screen/keyboard of the vision test system rather than the original control panel.

Given these objectives, the design philosophy evolved quite simply without many choices being required.

The stand-alone requirement indicated the use of micro-computer control and for reasons of supply, cost and general convenience the IBM Personal Computer (PC) was the obvious choice. The need to keep within the envelope of the machine meant that the PC had to be 'hidden' under the covers, this involved re-packaging the PC frame into a 19 inch rack format. A 'Cluster' adapter card provides the host communication link via a local area network (LAN).

The test accuracies to be met required more camera resolution than on the early tester and the cost constraints meant the choice of a CCTV unit rather than an 'automation' camera (the cheaper TV format cameras do not have square pixel arrays – in the one chosen 484 x 378 – in this particular application this is of no consequence). Camera control and image analysis, which for speed reasons has to be performed in hardware, was to be optimised for the application on a special card that would plug directly into an expansion slot on the PC; hence no requirement for bought-in control boxes or interface circuitry.

The mechanical layout of the test station was to add it to the end of the assembly track – a simple addition that would not interfere with the existing assembly and test points.

In the software, considerable effort was put into providing a completely integrated package that would be usable at all levels – by the machine operator, by the machine maintenance engineer, by the quality engineer and, via the host link, by production control. Good ergonomic presentation of the data on the screen and a substantial interactive 'help' library were essential parts of this package. Another important feature was to add a large number of continuously available running statistics on all aspects of the assembly machine's operation, this gives the machine minder a much deeper insight into how the machine is performing.

5. THE SOLUTION – GENERAL MECHANICS

Figure 4 is a view of a completed assembly machine with the vision station and the PC controller fitted, Figure 5 is a close-up view of the vision station.

The assembly machine feeds springs and pads from two bowl feeders into the spring insertion station, completed parts then pass along a linear vibratory buffer track to the buckle test station. An air slide moves individual parts into a precise location for the buckle test which is then operated from a cam-driven plunger. Parts were then, originally, ejected into an air jet pass/fail gate.

The vision station is inserted between the buckle tester and the pass/fail gate. Single assemblies are transported by air track from the buckle tester up to a stop-pin in front of the camera position; part presence is detected by an optical sensor and an air driven clamping mechanism is operated. This clamp has several functions:

a. A forked slide picks up two inclined surfaces on the pad, holding the pad down firmly on its datum surfaces against a reference surface in the track.

b. A compliant V-block on the slide LIGHTLY grips the spring against a back stop – this damps out vibrations in the spring and provides a repeatable location of the free end of the spring.

c. To further reduce part vibrations the track propulsion air is interrupted during the clamping cycle.

With the spring pad clamped, the PC controller issues a 'picture request' to the vision circuits and an image of the spring is captured and analysed. After processing the image and determining the height and orientation of the part, the stop-pin is withdrawn, the clamp released and the part passes to the pass/fail gate. The result of the previous buckle test is, of course, also an input to the pass criteria. All parts, including those that have failed the buckle test, are measured by the vision system to ensure that the statistical records are unbiased.

Particular care was taken with the mechanical design to keep all parts of the camera station very rigid, for example, the lens and the camera body are individually held to the base block to avoid reliance on the lens mounting ring.

6. THE SOLUTION – VISION SYSTEM

Both speed and financial constraints determined that there was no readily available 'off-the-shelf' commercial vision system that could do the job. Thus there was freedom to design a totally optimised system; the need to produce over thirty installations made this even more economic. The controller, as mentioned previously, had been chosen to be an IBM PC, therefore it was desirable that all of the vision electronics should be packaged on a single PC card.

From a study of the tolerances required on the tests it was clear that a camera of more than 256 pixels square was needed – good quality, small, reasonably priced units are the TM-34/36 series from the Pulnix Company; the 36RV version was chosen. This is a 625 line CCIR standard camera with external clock drive, the usable picture area is 484 lines of 378 pixels each. The light source was chosen to be an infra-red diode – this had the advantages of extreme cheapness and small size coupled with good life and the ability to use an infra red acceptance filter to minimise stray lighting effects.

Since the part is so small it is necessary to work at a real magnification of approximately X2, to achieve this with a standard lens it was necessary to use the technique of mounting the lens 'back-to-front'. The light source diode uses a diffusing element followed by a small doublet condenser lens to produce a uniform back lighting source of about 10mm diameter.

The output of the Pulnix camera is an RS170 composite TV signal. However, as the system externally synchronises the camera from the PC vision card, there is no requirement for the video synchronising pulses. The 'front end' of the card contains analogue circuits, these establish both black and white references for the signal and provide binary thresholding at a selectable percentage of the picture contrast; all of the reference levels track with the average picture level so that the binary image is very little affected by absolute light level or lens stop. The binary image is passed into the analysing logic, this logic drastically reduces the data content of the image by finding the 'first encountered vertical edge' and stores this in a small area of RAM that is memory mapped into the PC's address space.

The card is not a general vision processor, in that it only does this one specific image transformation; furthermore, for any application it does imply some prior knowledge of the image characteristics – this is not a problem with most industrial applications and it does have the great advantage of permitting the card to process images at camera frame rate.

Figure 6 shows a sequence of images illustrating the image processing algorithm; 6a is the greyscale image at the camera output, 6b is the binary image taken at about 70% threshold, 6c and 6d show the tolerance of the binary image to camera f-stop. 6e shows the processed image after the application of the 'first edge' algorithm. Note that the algorithm ignores reentrant areas and image holes, only the left-most boundary is retained.

After a 'picture request', the data returned to the PC in the memory mapped area is a string of just 484 numbers – the pixel counts along each picture line of the processed image's boundary. The edge finding logic includes digital filtering, arranged to eliminate spurious one or two pixel noise pulses. Two programmable 'windows' are available, allowing masking of unwanted objects or areas and also providing a default baseline for the edge algorithm (use of a baseline can be seen in Figure 6e).

From the initiation of a 'picture request' signal the worst-case time for the analysed data to be ready is 60 milliseconds, the average time is 50 milliseconds.

7. THE SOLUTION - SOFTWARE PACKAGE

With the addition of a PC to drive the vision card, it was clearly sensible to maximise the use of this computing power and transfer the whole operator control of the assembly machine to the PC. The keyboard and screen were mounted on a stalk in place of the machine's original control panel (see Figure 4).

An IBM utility program, ADMS (Application Display Management System), was used to generate the operator interface. This program package provides an easy way to format input and output screens, menus etc. A variety of screens were designed for the normal running, for set-up, for calibration and for machine maintenance. Figure 7 shows the normal running screen, Figure 8 the calibration input screen and Figure 9 one of the many inter-active 'help' screens that have been provided.

As can be seen, the run-time screen gives the operator a concise view of the machine's statistical operation. Details are shown (upper right) of the height and orientation measurements for both the whole machine product population and for a short-term running sample (the timescales of each being programmable). Failures are analysed by type (lower left) and there is a detailed analysis of the vision system's performance (lower right). These statistics are recorded continuously and are transmitted, via the 'Cluster' adapter and LAN, to a master PC which is itself linked to the Greenock plant's mainframe computer.

The orientation measurement is intrinsically a geometrical calculation so no calibration is needed, however, for the height test there is a calibration routine; this uses parts of a known height. These are fed individually into the vision station and subsequently measured with a gauge - the gauge measurements are keyed into the calibration screen (Figure 8) and the calibration routines automatically produce a statistically optimised setting of the image magnification and zero offset to be used for subsequent parts.

Extensive 'help' routines are available for all of the machine's activities, Figure 9 shows one example. These routines have several nested levels of detail accessed by highlighted keywords. The aim of this design is that all machine operations should be possible without reference to any written manuals or documents.

The application routines for the majority of the analysis and control are written in 'C' with some small sections in Assembler (the Assembler modules are for convenience in handling some of the machine's hardware interface, rather than for speed). The image analysis routine can be summarised as follows:

a. The image is checked to see that the spring is wholly within the picture - this just requires testing for the presence of a certain number of baseline points at the top and bottom of the image.

b. The edges of the spring are found by observing the points at which there are sufficiently high-slope deviations from the baseline. The difference of these two points (i.e. the spring diameter) is checked for 'reasonableness'.

c. The position of the 'step' produced by the cut end of the spring is found by using an incremental 'worm' that scans the image contour between the two edges looking for a step change in a locally integrated value combined with a slope change.

d. The height is found by a local average taken at a point directly before the step.

8. EXPERIENCE AND CONCLUSIONS

All of the 32 assembly machines were converted by the summer of 1986. Overall machine production rate is a little over one and a half seconds per part, so the Greenock plant now has the capability of making some 2 million, 100% vision inspected, parts per week!

The installations finally cost the plant around £7500 per machine - slightly more than the target due to some added function - but still far below the sort of price that is customarily asked for vision systems.

Presently the vision system is acting simply as an inspector, however, it is clearly possible to close a loop round the whole machine from the statistics within the PC to the adjustment mechanism of the spring insertion station to make the machine completely automatic. It is also clear that this type of PC based vision system could have many other applications where low cost, high speed and a fairly simple part is involved.

9. UK DEVELOPMENT/MANUFACTURING PROCESS CENTRE DESIGN TEAM

This system was developed in the UK DMPC department at IBM's Hursley Laboratory (near Winchester) in the second half of 1985 by a team of three engineers:

Maurice Bosier, mechanical design. Graham Caulfield, software design. Patrick O'Neill, vision card design.

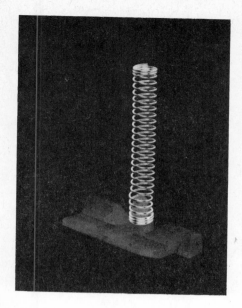

Fig 1   The spring pad assembly

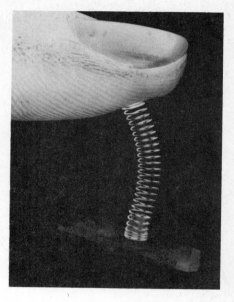

Fig 2   The spring pad assembly

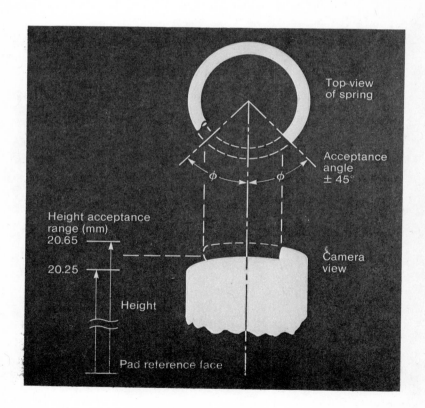

Fig 3   The required tests

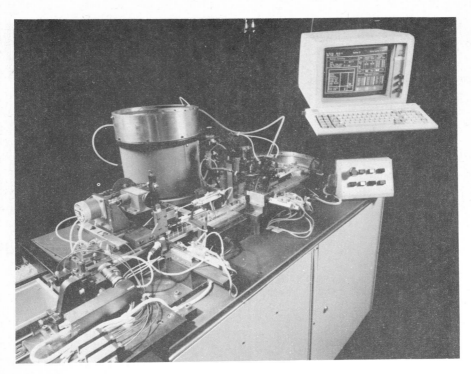

Fig 4   Overall view of the spring pad assembly and test machine.
The central station is the functional 'buckle test':
lower left — the vision inspection station
upper right — the assembly station with its two bowl feeders

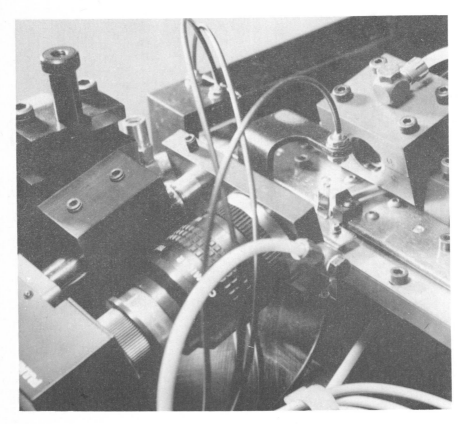

Fig 5   Close-up of the vision station. A spring pad is clamped at the
inspection station:
upper right — infra-red lighting source
lower left — camera and lens

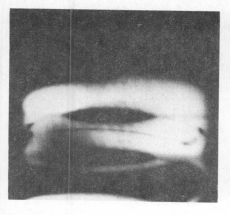

Fig 6a   Grey-scale image

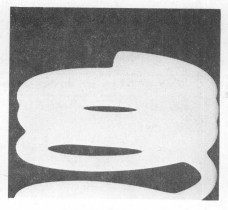

Fig 6b   Binary image at F/8

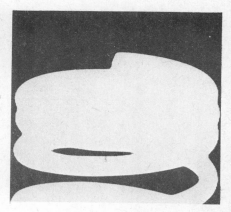

Fig 6c   Binary image at F/4

Fig 6d   Binary image at F/16 (note the tolerance to light level)

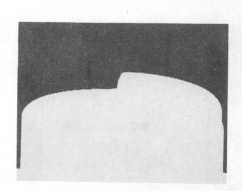

Fig 6e   Processed image showing the 'first edge' algorithm

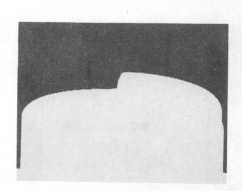

Fig 7   The normal run-mode screen (Figs 7, 8 and 9 are negative images from a colour display)

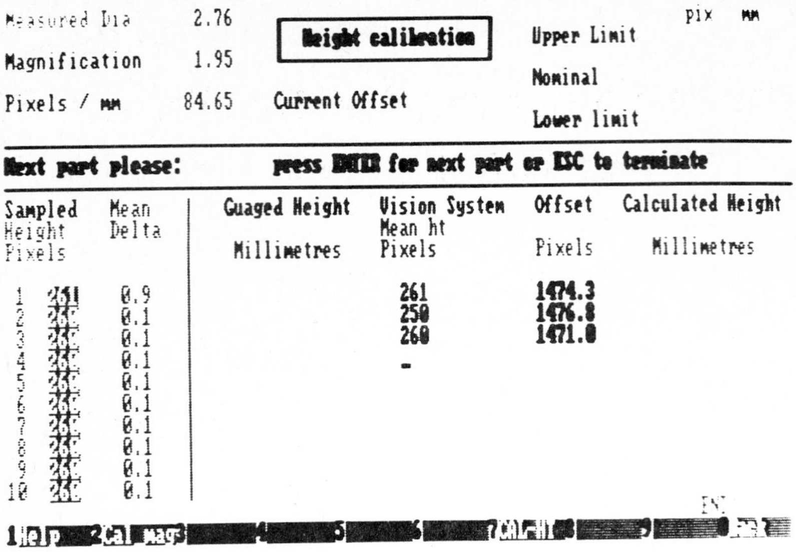

Measured Dia	2.76		Height calibration	Upper Limit	pix MM
Magnification	1.95			Nominal	
Pixels / MM	84.65		Current Offset	Lower limit	

Next part please: press ENTER for next part or ESC to terminate

Sampled Height Pixels	Mean Delta	Guaged Height Millimetres	Vision System Mean ht Pixels	Offset Pixels	Calculated Height Millimetres
1 251	0.9		261	1474.3	
2	0.1		250	1476.8	
3	0.1		260	1471.0	
4	0.1		-		
5	0.1				
6	0.1				
7	0.1				
8	0.1				
9	0.1				
10	0.1				

1 Help 2 Cal Mag

Fig 8   The calibration screen

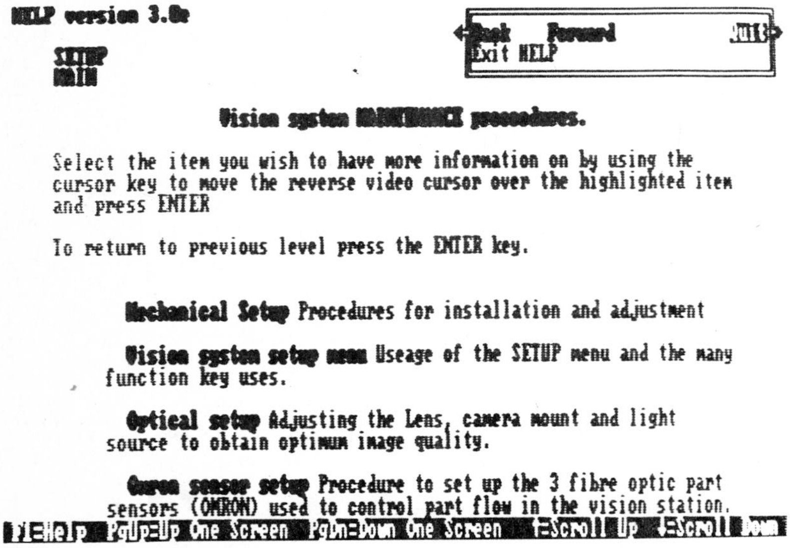

HELP version 3.0e

SETUP
MAIN

Back   Forward
Exit HELP

Vision system MAINTENANCE procedures.

Select the item you wish to have more information on by using the cursor key to move the reverse video cursor over the highlighted item and press ENTER

To return to previous level press the ENTER key.

Mechanical Setup Procedures for installation and adjustment

Vision system setup menu Useage of the SETUP menu and the many function key uses.

Optical setup Adjusting the Lens, camera mount and light source to obtain optimum image quality.

Curen sensor setup Procedure to set up the 3 fibre optic part sensors (OMRON) used to control part flow in the vision station.

1=Help   PgUp=Up One Screen   PgDn=Down One Screen   f=Scroll Up   f=Scroll Down

Fig 9   A typical help screen

C360/86

Programmable fixturing for flexible manufacturing systems

H C LINTON, BSc, **J B C DAVIES**, BSc, MSc and **T G J MUIR**, BSc
Department of Mechanical Engineering, Heriot-Watt University, Edinburgh

SYNOPSIS   This paper describes the development of a system which facilitates the programmable fixturing of a family of castings. It discusses previous work done in the area and categorizes two approaches to the problem. The adopted approach exploits the similarities between workpieces enabling the design of an Adjustable Integral Fixture Pallet. A computer controlled fixturing cell incorporating a Pallet Loading Station, a vision sensor and a robot is described. Future developments are also discussed.

1 INTRODUCTION

1.1 Principles of Fixturing

In all metal cutting processes, it is imperative to ensure a known relationship between workpiece and cutting tool(s). For rotational parts, in addition to providing accurate location, the self-centering chuck device can be used to cater for wide variation in workpiece size. Prismatic workpieces, however, demand auxiliary tooling whose function is to facilitate accurate location and ensure stability throughout the machining cycle. In many instances, the design of the FIXTURE may require as much thought as the design of the workpiece itself. This is particularly true if the workpieces are complex in form.

The fixture design process can be separated into two phases (1). The initial concern of the fixture designer is establishing the most suitable orientation with respect to the tool. Primary considerations during this phase will be satisfying stability requirements, assessment of the accuracy of location points and the magnitude of the cutting forces to be applied to the workpiece. The second phase attempts to eliminate the remaining freedom of motion. The 3-2-1 location principle will feature heavily during this phase of the fixture design and the designer will make every effort to achieve this ideal. Elimination of the remaining degrees freedom is the task of strategically positioned clamps.

It is frequently necessary to pre-machine location features in order to establish the known relationship between the workpiece and machine spindle. In these instances, two workholding devices may be required.

1.2 The use of Pallets in Flexible Manufacturing Systems

It has become accepted practice in manufacturing industry that prismatic workpieces are loaded onto pallets (fig.1). These pallets, coupled with their associated fixturing systems are adopted to fulfil the normal workholding and location functions. In addition, however, workholding pallets simplify the problems of automating the manipulation and transportation of irregularly shaped workpieces. Easing the difficulties of workpiece transportation simply relocates the problem and imposes strict discipline in Pallet Load and Unload areas of Flexible Manufacturing Systems (FMS). It is here that a substantial amount of human intervention can be found. Traditionally undertaken by 'unskilled labour', the human operator requires only a short period of time to learn how to perform such tasks. This work, however, is usually boring, often strenuous and sometimes hazardous. Moreover, humans are ill-suited to working as part of a system which is paced by automatic machinery. As yet, there does not exist a machine(s) which can automatically pick up an irregularly shaped workpiece, place it correctly in a fixture and clamp the workpiece.

The automatic clamping of irregularly shaped workpieces demand the emergence of a new generation of machine, simple in design and operation, yet of sufficient 'intelligence' to justify their substitution for the human worker. This paper describes work at Heriot-Watt University aimed at providing such facilities. It is the opinion of the authors that this work will improve the likelihood of unmanned flexible manufacture.

2 AUTOMATING THE FIXTURING PROCESS

More and more users of advanced manufacturing systems are expressing an interest in fully automated, unattended loading and fixturing of parts onto pallets (2). This need has been partially satisfied by the emergence of power clamping, utilising air or hydraulic power sources that are brought to a machine tool (3). Power vices, clamping heads, indexing clamps and supports units (4), can be combined to provide automatically activated fixturing of specific workpieces. These devices can also be conveniently activated by a host Numerical (NC) or Computer Numerical Controller (CNC).

The application of these units has provided remarkable payoffs. In addition to providing time savings in machine setup, their use has made significant contributions to increases in productivity, improved quality and increased tool life.

The nature of these devices, however, renders the accommodation of large variations in workpiece geometry a difficult task. Their operation is such that they become single-purpose tools.

Work done by Asada et al (5), demonstrates the use of a robot to configure standard fixturing elements to accommodate varying workpiece geometry. The approach incorporates kinematic analysis of the fixturing problem and the use of a CAD (Computer-Aided-Design) database description of the workpiece is an attractive feature of the work. The methods and analysis adopted, however, have focussed on solving the problems related to automatic assembly. The criteria for solving such problems differ in many ways from those in metal cutting.

Later work by Asada, extended the application of his system to sheet metal parts (6). This system made use of the designer's expertise in conjunction with CAD surface data for absolute determination of the fixture configuration.

Asada's ideas of assembling fixtures from standard elements is an attractive solution to the problem. Modular fixturing systems enabling manual assembly of workholding devices in this way have been commercially available for several years (2).

Neads et al (11) adopts the approach of assembling fixturing elements in a similar fashion to Asada. Initial efforts in this work however, concentrated on reducing the number of elements necessary to satisfy the fixturing needs of 2½ dimensional workpieces. This has been achieved by the development of washers and couplings which provide varying height and angular orientations required to uniquely define the location of a workpiece on a platen containing a matrix of tapped holes. A specially designed turret device operates in conjunction with an X-Y table under computer control providing the ability to assemble the fixturing elements according to demand.

In contrast, Cutkosky et al (7) showed how 'moveable plungers' could be pre-configured to accommodate the programmable clamping of turbine blades. Cutkosky continued to show how the moveable plungers could be adjusted according to 'master profiles' held within a computer database. This facility provided the required fixture flexibility necessary for the holding of blades of differing shapes.

These examples of flexible fixturing, illustrate the two schools of thought in the application of automation to workholding:-

1. The assembly of fixture devices from standard elements, using a form of robotic machine producing a SINGLE-PURPOSE fixture.

2. Designing a degree of adaptability into the fixture resulting in a MULTI-PURPOSE workholding device.

The work done by the authors falls into the second of these two categories.

3 THE GROUP TECHNOLOGY APPROACH TO FLEXIBLE FIXTURING

3.1 Component Definition

GROUP TECHNOLOGY has been with us for many years. The technique for bringing together related or similar components in a production process has proved successful in the business of small batch manufacture. By exploiting the similarities which are known to exist among populations of components, group technology attempts to reduce the total cost of piece-part manufacture. Cells are created to manufacture types and size ranges of workpieces. Groups of machines chosen for each FAMILY are situated together in a group layout.

The crux of the problem of introducing group technology, is the identification of the component families. Considerable emphasis has been placed on this aspect. In particular, with regard to component classification and other techniques which facilitate this sorting process.

Workpiece grouping forms the basis of flexible manufacture. Components require classification into two basic categories relating to their manufacturing and automatic handling needs (8). Further classification should take account of workpiece geometry, size, weight, material, accuracy and finish, initial form and batch size. All these items influence the capacity, tooling configurations and workholding methods and devices employed in the manufacture of the component group.

Figure 2 shows a component GROUP which would typify a family of workpieces under the basic classification criteria discussed. Their geometric similarity is the most dominating common feature of the group. It is this feature which is the largest contributor to the ability to develop common fixturing.

3.2 Pallet Development

Figure 3a shows the adaptation of a classical clamping technique in order to provide flexibility in clamping. Two features give the arrangement its ability to adjust; the LEADSCREW for height adjustment and the SLIDING BLOCK for position. In addition to providing height adjustment, the leadscrew applies the central force required for workpiece stability during the machining process. CLAMPING HEADS act as simply supported beams; the workpiece and STEPPED SUPPORT BLOCKS supplying the required support.

The locus of clamp movement is directly related to the workpiece geometry and the machining operations to be carried out. Sites on each of the workpieces in the family are chosen as the most favourable clamping regions. The locus of these sites, approximated to straight lines, has determined the optimum clamping locii. Figure 3b shows how this is achieved.

The adjustable Integral Fixture Pallet, figure 4, makes use of three point clamping. The position of the clamping elements on each of the axes, however, can be independently positioned and it is feasible that the pallet will enable the fixturing of a wider group of workpieces.

Throughout the pallet development phase, a concious effort was made to design a purely mechanical clamping device. All power would be provided by devices operating on the pallet allowing the fixturing cell concept to fall in line with present day materials handling technology for FMS. The pallet would, therefore, be free from connection to any external power services.

4 THE DEVELOPMENT OF A PROGRAMMABLE SYSTEM

4.1 Pallet Loading Station

The adjustable nature of the integral fixture pallet,

lends itself to automatic clamp manipulation. The PALLET LOADING STATION (PLS) is a machine currently under development which locates the pallet and allows its configuration to be determined under software control.

The machine comprises 'discrete position' actuators which are powered by pneumatics. Proximity sensors, which operate as normally open switches are activated whenever the magnetic piston of the actuator is within its range. The proximity sensors therefore, are used to determine the clamping positions and can be manually adjusted provided certain simple rules of control are obeyed (12). Sensor signals are fed back to a specially developed PLS electronic interface which is controlled by microcomputer. The interface provides the actuator valves with the correct drive signals depending on the previously selected position and also monitors the progress of the actuator as it approaches the software selected sensor. Presently, the machine provides three-axis control accommodating the manipulation of the clamping elements which form an integral part of the pallet.

Specially developed locators mounted on each of the pallet sliding blocks, can form a detachable link with the positioners. These locators primarily enable the pallets to be detached from the PLS once the workpiece has been located.

Clamp Transport and Adjustment Mechanisms (CTAM) provide the pallet leadscrews with the rotary drive required for height adjustment and positive location of the parts. A splined drive shaft is elevated into the base of the pallet leadscrew by a specially designed diaphram mechanism that provides adequate thrust in a suitably compact volume. Rotary motion can then be transmitted from a pneumatic motor, via the CTAMS, to the pallet leadscrews, providing a clamping torque in the order of 153 Nm (114 ft. lb) which is sufficient to locate the workpieces.

The complete clamp adjustment mechanism is to be rigidly attached to the pneumatic actuators to ensure reliable, positive alignment of the rotary driving mechanism with the leadscrew.

All PLS movements are controlled by a microcomputer. The PLS interface allows the initialisation of the pallet clamping elements, irrespective of their starting position. The controlling software currently enables user defined clamping configurations for the pallet, simulating the recognition of a particular workpiece within the defined set.

Software developments have also included an interactive command controller incorporating an offline program definition module.

4.2 Component Recognition

Automatic configuration of the pallet is dependent upon a recognition process. This process can be performed in many ways. The parametric nature of the parts enables the identification of 'tell-tale' dimensions and features. It is important, however, that the recognition process attempts to replace some of the sub-conscious human inspection lost when automating the fixturing function. The failure of cores may result in casting features being deformed or missing; surface depressions and blow holes are defects easily detected by the human operator. The determination of workpiece acceptability is, therefore,

an important feature of the recognition system.

A vision system has been utilised to determine the component being handled and to provide the 'visual' inspection required. The vision sensor, based on a DYNAMIC RAM chip, can distinguish between the components in the set by examination of the two dimensional silhouette image. Image parameters which relate to area, perimeter, length and breadth of the current workpiece are compared with data extracted during a learning phase. (12),(13)

A lighting box illuminates the workpiece enhancing the quality of the image. The vision sensor, controlled by microcomputer, condenses the image data into a single parameter revealing the workpiece currently being processed.

4.3 In-Cell Transportation

A UNIMATE 2000B turret robot utilised within the cell, has sufficient capacity to handle all the workpieces in the set. A specially designed gripper allows the robot to grasp all of the workpieces by a common feature. The nature of the components and gripper, demands unique pick and place positions. The controlling microcomputer, however, allows the selection of programmed positions in accordance with the data relayed from the recognition system.

4.4 The Complete System

Figure 4 illustrates the proposed system and highlights the areas that are currently in operation. Each of the elements within the system form individual modules with the ability to operate as such. A computer network, however, allows the communication of data between each of the controlling computers. Presently, the vision system performs the recognition function and subsequently passes the required information to the PLS and robot.

The robot, contains within its memory, sets of pre-programmed pick and place instructions which allows the reliable manipulation of the workpieces. On receipt of the workpiece information from the vision sensor, the robot computer increments the program steps in order to select the correct portion of the robot memory. The robot is then activated and the workpiece is manipulated onto the pallet.

The vision information is also provided to the PLS controlling computer. On completion of the manipulation task, the robot computer instructs a supervisory computer that the workpiece is ready for clamping. The supervisory computer acknowledges the completion of the manipulation and task and then to instructs the PLS to clamp the workpiece. The cell software is still in the early stages of development.

Initial attempts at simple network software have provided promising results. This software however, will be the subject of further development in the future. At present, the supervisory computer simply orchestrates the operation of the constituent elements of the cell.

5 FUTURE WORK

5.1 Casting Inspection

It is intended that the automatic fixturing cell will have the ability to eliminate pre-machining of the group of castings being handled. These operations have often been a necessary part of the setting process. The establishment of datums ensuring that all faces and features to be machined will 'clean up', that metal thicknesses are correctly balanced, are vital in ensuring the quality of subsequent machining operations. In light of the advancements made in inspection, it is possible to establish an optimum datum using an inspection machine to carry out a series of programmed moves on an unmachined workpiece. Work is being undertaken to develop an inspection machine, capable of operating within the confines of the PLS and establishing the required linear and angular offsets for each casting to ensure the quality of the machined part.

The inspection machine, fig. 7, is a simple three-axis device, incorporating a three dimensional triggering probe and controlled by microcomputer. Measurement of each casting will be performed in a sequence dependent upon the previously defined critical dimensions. The machine will be positioned to an accuracy of $\pm$ 0.2mm, sufficient to quantify casting characteristics but not finished components.

It is envisaged that the necessary component offset data will be communicated directly to the machine tool controller.

5.3 Expanding the acceptable component set

The current Programmable Fixturing cell is limited to a small set of components. This is primarily due to the initial constraints imposed on the design of the pallet. Work is currently being undertaken to expand the component range and particularly to the examination of component groups with similar properties to those used in this work. New pallets are being designed to accommodate the fixturing of different workpiece geometries. These pallets, however, are to be such that they can be manipulated by the current PLS.

Future work will examine the flexibility of the PLS with the aim of developing a more universally applicable programmable fixturing machine.

ACKNOWLEDGEMENTS

The authors would like to acknowledge colleagues in the Department of Mechanical Engineering and in particular Mr G.J. Smith and Mr N. Lunan. Our thanks extend to the National Engineering Laboratories, East Kilbride for their support during this work.

6 REFERENCES

1. Ferreiara, P.M., Kochar, B., Liu, C.R., Chandru, V., Afix: An Expert System Approach to Fixture Design Computer-Aided/Intelligent Process Planning, ASME PED - Vol. 19 1985, pp. 73-82.

2. Quinlan, C., New Ideas in cost cutting, fast changing fixturing, Tooling and Production (USA), 1984, Vol. 1, Pt. 50, pp. 44-48

3. For Fast Clamping Setups - take the power to the madhine, Manufacturing Engineering (USA), March 1977, pp. 46-47.

4. Hydrajaws - FMS Pallet Clamping Systems, Technical Brochure, Hydrajaws Ltd., Birmingham, U.K.

5. Asada, H., Andrew, B., Kinematic Analysis and Design for Automatic Workpart Fixturing in Flexible Assembly, 2nd International Symposium of Robotics Research, Kyoto, Japan, August 1984.

6. Asada, H., Fields, A., Design of Flexible Fixtures Reconfigured by Robot Manipulators, Robotics and Manufacturing Automation, ASME PED - Vol. 15 1985, pp. 251-257.

7. Cutkosky, M.R., Kurokawa, E., Wright, P.K., Programmable Conformable Clamps, Autofact 4 Conference, Philadelphia, Pennsylvania, Society of Manufacturing Engineers, Nov. 1982.

8. Christie, N.R., Knight, J.A.G., Lebrecht, H.M., Marks, J.F., Scott, W., Machine Tool Design, Automated Small-Batch Production, ASP Technical Study, Department of Industry, 1978, pp. 65-135.

9. Ogden, H., Monitoring, Gauging and Inspection, Automated Small-Batch Production, ASP Technical Study, Department of Industry, 1978, pp. 137-152.

10. Linton, H.C., Davies, J.B.C., Flexible Clamping for a New Programmable Prismatic Workpiece Palletizing System, Heriot-Watt University Internal Report.

11. Neads, S.J. Graham, D. and Woodwork, J.R., Toward a fixture building robot, Proceedings of I.Mech.E. Conference on Robotics, London, Dec. 1984, pp 99 - 104.

12. Davies, J.B.C., Linton, H.C., Automated Prismatic Component Recognition and Clamping System. Proc. 3rd International Conference on Automated Manufacture, IFS Summit, NEC Birmingham, U.K.

13. Someron, A.V., The Snap Camera Manual Provisional Copy, Sept. 1984.

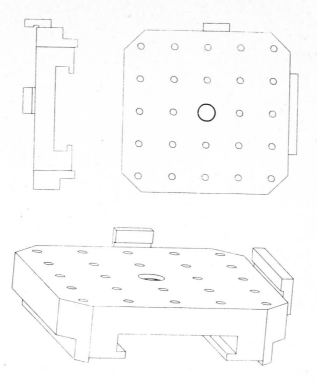

Fig 1   Typical component pallet with a matrix of holes and registers
          for fixtures

Fig 2   Component group

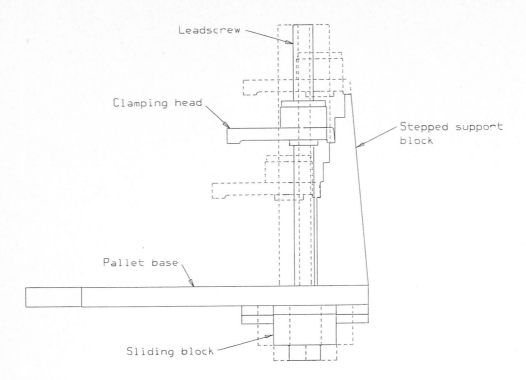

Fig 3a   The adaptation of a classical clamping technique

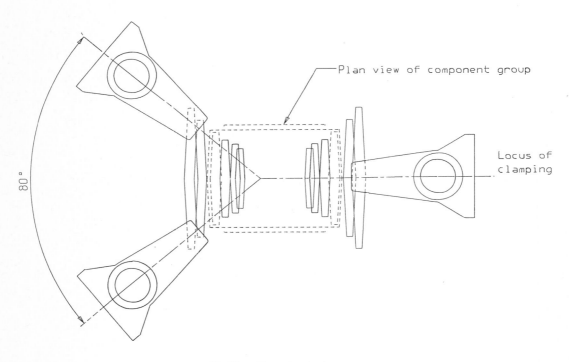

Fig 3b   Clamping loci

Fig 4   Adjustable integral fixture pallet with largest of component
group clamped

Fig 5   Pallet loading station

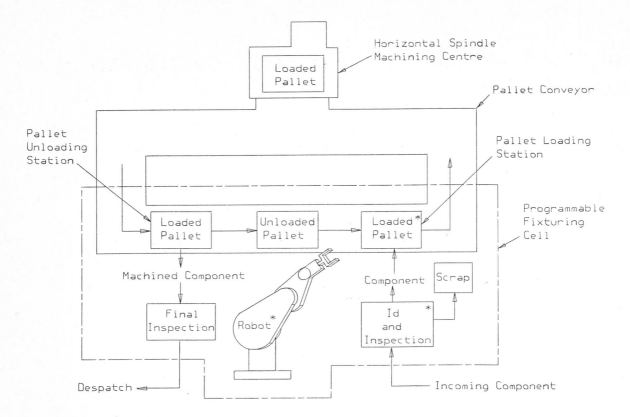

Fig 6   Programmable fixturing cell (* shows modules currently in operation)

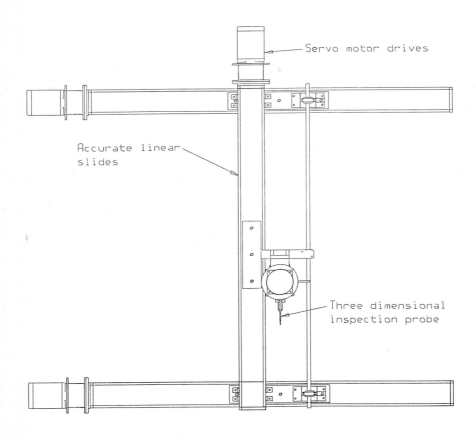

Fig 7   Inspection system

C365/86

Robotics handling of composite materials in a lay-up cell

E J WRIGHT, BSc, PhD, CEng, MIMechE, **J P J CURRAN**, BSc, PhD and **P J ARMSTRONG**, BSc, PhD, CEng, MIMechE
Department of Mechanical and Industrial Engineering, Queen's University of Belfast, Northern Ireland

SYNOPSIS   The operations involved in lay-up of composite prepreg materials are reviewed and a design of adaptive vacuum gripper to pick up cut plies of arbitrary shape is described. Initial test results confirm an acceptable performance for the gripper. A structure for a lay-up development cell with robotic handling of prepreg plies is proposed and a form of sensory control is described. The vacuum gripper is used with a Cincinnati T^3 hydraulic robot, a two camera vision system and off-line programming to demonstrate stacking of cut plies in the presence of positional errors. The success of this stacking operation is examined.

1   INTRODUCTION

A comprehensive research study into the use of robots in the automation of manufacturing operations for composite materials is being conducted at Queen's University of Belfast in collaboration with Short Bros, plc. The work is sponsored by the ACME Directorate. The overall research programme comprises several individual projects concerned with the principal operations used in small batch manufacture of aircraft components. These include the lay-up of material in a mould prior to curing in an autoclave, drilling and trimming of cured components and riveting of sub-assemblies in preparation for final aircraft assembly. An early phase of the work also examined the use of robots in the inspection of cured components. This led to the development by Short Bros, of a robot-based inspection system for composite manufacture. Several aspects of the research programme concerning drilling and inspection operations and a robot off-line programming procedure have been reported previously (1, 2, 3). The present paper concentrates on the use of a robot in lay-up operations of composite prepreg broadgoods. This is one of the available forms of composite materials in which reinforcing fibres are woven into a fabric and preimpregnated with a resin matrix. The uncured fabric is flexible and is normally stored in a roll at low temperature until required for use.

In outline, the manufacture of components from composite prepreg materials involves assembly in a lay-up mould of a number of layers of the material (called plies) cut from a roll, with the fibres oriented to exploit their considerable strength. Air is removed from the mould to prevent voids occuring in the component and the lay-up assembly is subjected to a temperature and pressure curing programme in an autoclave to produce the final, rigid component.

In manufacturing terms, the process involves several distinct operations;

dispensing the prepreg from a roll onto the cutting table,

cutting plies to shape,
lay-up in a component mould,
bagging the assembly of plies for debulking and curing,
curing in an autoclave.

Currently, most composite components for commercial aircraft are manufactured using manual methods for dispensing, cutting, lay-up and bagging, even though automatic dispensing systems and various forms of automatic cutting systems are available. This is mainly due to the small batch sizes and small range of components presently required. However, there is a general belief throughout the industry that demand will increase as more aircraft components are made from composite materials.

It has been shown that the manual lay-up process is a main contributor to composite component cost (4) and it is estimated that costs could be reduced by 25 - 50% through automation. In the U.S. aircraft industry, a number of studies have been made into automated composite manufacture (5, 6, 7, 8) aimed largely at military aircraft production. All include the use of robotic handling devices, either standard commercial robots or specially designed gantry robots to suit specific system requirements. Both lay-up of individual plies into the component mould and stacking of a number of plies on a flat surface prior to mould lay-up have been demonstrated and there seems to be a general conclusion that the individual ply lay-up method offers little improvement in quality or structural integrity of the component compared with lay-up of pre-assembled stacks. This is an important conclusion since not all component shapes lend themselves to automatic lay-up in the mould. In such cases, robotic stacking of plies could be used in conjunction with final manual lay-up in the component mould.

Short Bros, as part of their development work associated with the research partnership have produced a simple vacuum gripper for lay-up of prepreg test pieces of fixed size and shape for material quality checks. One particularly useful feature of their work has been to automatically perform bagging of the test pieces lay-up both for debulking and in preparation for curing in an autoclave. In the present project, a prototype adaptive vacuum gripper has been developed to study the problems of robotic handling of more generally shaped composite pre-preg plies during the lay-up process. In addition, an experimental robot lay-up cell structure has been devised to investigate various degrees of automation of the lay-up process and its use in some preliminary tests on stacking of cut plies is described.

A concensus view has emerged across all of the projects in the research programmme of what seems the most appropriate way to utilize robots in small batch manufacture. This includes off-line programming of the robot tasks coupled with sensory control of the end effector. Sensory control enables the robot to cope with some degree of disorder in the manufacturing process. In addition, it offers a means for maintaining precision in the presence of inaccuracy intro-duced by the off-line programming process itself and by the robot mechanism.

There are two principal ways of utilizing such sensory information:-

1. To modify the stored robot program and hence correct the robot arm position.
2. To activate positioning drives within the end effector which correct the positional errors without changing the nominal position of the robot arm.

In principal, the first of these is the more appealing since it utilizes the joint drives of the robot to implement corrective motions, resulting in a simpler design of end effector. This is the approach used in the present work and is described in more detail later. The second method duplicates some of the positioning drives of the robot, but only for a considerably smaller range of motions. Although this may appear an unnecessary duplication, it does offer the prospect of achieving a better positioning resolution than that available from the robot itself. However, the added weight and bulk may be difficult to accommodate in certain appli-cations.

2 ADAPTIVE VACUUM GRIPPER

A prototype vacuum gripper has been developed to pick up plies of random size and shape from a cutting table where they have been cut in a known nested pattern. Each ply needs to be picked up without disturbing others adjacent to it and without incurring damage. It is sub-sequently deposited either in a flat stack for later insertion in the component mould or directly in the component mould. Only the flat stacking operation is considered here. In general, the gripper head must be lightweight to accommodate the limited lifting capacity of a robot and to allow the use of a large head for large component lay-up.

The head structure used to meet these design requirements contains a number of cells within which air pressure or vacuum can be individually controlled. All of the pneumatic valves for con-trolling cell pressure are contained in a separate rack assembly and connected to the gripper head through flexible plastic tubing. The proto-type experimental gripper has 100 cells in the gripper head each of size 50mm x 50mm. Each cell has one pressure connection on its back face and four small holes on the front face to bring the ply in contact with the cell pressure or vacuum level. The front face of the cell is flat but it can be fitted with a profiled porous inter-face element to allow lay-up directly into a com-ponent mould. A schematic diagram of the func-tional elements of the experimental gripper is shown in Fig. 1 and a photograph of the gripper head is shown in Fig. 2. The air pressure and vacuum supplies are derived from a filtered shop air source providing a pressure of around 550 kPa. Vacuum is generated using a pair of multi-ejector vacuum generators operating in parallel and providing a free air capacity of 500 litres per min. at pressures down to -55 kPa.

The adaptive vacuum gripper obtains its ability to adapt to different ply shapes and sizes through a microcomputer controller which applies a pattern of vacuum and pressure throughout the cells to match each particular ply. The vacuum is used to pick up the ply and the surrounding positive pressure to ensure that adjacent plies remain undisturbed. In the experimental gripper, this pattern can be selected interactively by an operator at a terminal or be determined by a separate supervisory computer which is co-ordinating the overall lay-up process.

For interactive use, a special microcomputer display was devised as a gripper programming aid. An example of this is shown in Fig. 3. In addition to the options menu and screen space for dialogue between the operator and the controller, there is a status display of all 100 valves supplying the head cells, the two main valves providing pressure or vacuum to the head and a diagrammatic view of a simple gantry manipulator. This manipulator was built as a substitute for the T^3 robot which is installed at Short Bros, to allow early work with the gripper to be per-formed at the University.

The network link shown in Fig. 1 was not used in the work described here. However, in a production version of the gripper it would be available for all external communications in place of the several dedicated links shown in the diagram. In addition, the general purpose microcomputer would be replaced by a single board controller.

2.1 Gripper Performance Tests

The decisions to remove the pressure control valves from the gripper head and mount them on a separate rack with flexible tubing connections raised questions of the effects this would have on pressures available at the head and the speed of response of the pneumatic system. Preliminary tests with a small four cell gripper had shown that a vacuum level of around -10 kPa was suffi-cient to pick up a ply. Levels greater than -60 kPa could result in damage to the ply through distortion at the holes in the gripper face. How-ever, the dynamic effects of the necessarily long connecting tubing (8m long for use with the T^3 robot) was uncertain.

A set of tests were run to measure the performance of the experimental gripper. Some of the results from these tests are shown in Figs. 4 and 5. The tests were conducted using 4mm diameter connecting tubing to each cell. A piece of prepreg which covered half of the gripper surface was used as the ply and a set of 6 pressure transducers were installed to measure cell pressures. The locations of these are shown in Fig. 4.

The results shown in Fig. 4 are pressures in the instrumented cells for the case of pick up of the ply. Prior to pick up, the cells have been at atmospheric pressure (zero gauge pressure) and the cell control valves have been set to reflect the shape of the ply. The starting time for these plots occurs at the instant the controller command is issued to apply vacuum to the chosen cells. A delay of 66 msec, occurs before the cell pressures start to fall. This reflects the length of the connecting tubing (23.5 msec, delay for 8 m tubing and an acoustic velocity of 340 m/s) and the response of the vacuum supply valve and ejector system to the controller command. Thereafter, the pressure drops initially quite quickly and then at a decreasing rate. The pick up pressure of -10 kPa is reached after 160 msec. The cell pressures continue to fall, arriving at a more or less steady level of around -33 kPa in less than one second from the command. Clearly this response is acceptable for most robotic applications.

The results shown in Fig. 5 are for the case where the ply is being deposited at the lay-up point. Again, the time scale is synchronised with the control command to change from applying vacuum to applying positive pressure at the cells covered by the ply. However, in this case both the main supply valves and each of the active cell valves are switched by the command. A delay of 47 msec, occurs before the arrival of a positive pressure wave which rapidly removes the cell vacuum pressures. These results also illustrate a further useful feature of the gripper, an adaptive release pressure. Fresh composite prepreg is quite tacky at room temperature and shows a tendency to adhere to the surface in contact with it. In Fig. 5, this has occurred in the region of cells 4, 5 and 6. However, the resulting reduction in airflow out of the cells causes the pressure to continue to rise until such time as the adhesion is broken and the cell pressure falls rapidly. In Fig. 5, separation is complete after 630 msec. The remaining positive pressure of 4 kPa in the cells results from the continued proximity of the gripper to the ply, producing an air separation layer. Again, this release time should present no significant problems for the robotic system. The rapid change over from full vacuum to positive pressure levels shown in this figure offers the prospect of using a modulated lay-up of the ply in which release is controlled, say, from one end to the other. This should assist lay-up directly into a component mould where it is essential to avoid trapping pockets of air between successive plies.

Operating the gripper with all cells uncovered produces cell pressures of only about -0.5 kPa with vacuum applied and 2 kPa with positive pressures applied. Thus, significant cell pressure differences are seen to exist during various stages of the ply handling cycle. This suggests that if each cell were fitted with a simple binary pressure sensor then the gripper could also act as a sensor to indicate the presence or otherwise of a ply. In addition, with careful attention to the porting arrangements for each cell, this may also prove useful for locating the position and orientation of a ply on the gripper surface.

3 EXPERIMENTAL LAY-UP CELL

Although the vacuum gripper described in the previous sections can be used simply as an intelligent end effector for stand alone robotic applications, its intended use is as part of an automated lay-up cell. Ideally, such a cell would be capable of fully automatic operation including dispensing of the prepreg with automatic removal of the protective backing sheet, cutting the prepreg into plies, pick up and stacking of plies or direct lay-up in the component mould and finally bagging the assembly for debulking and curing. However, such an ideal is unlikely to be feasible for all possible component geometries. A more realistic cell would include some manual operations to supplement those which can be readily performed by one or more robots.

An experimental lay-up cell structure has been devised to enable investigations into various degrees of automation. Fig. 6 shows a diagrammatic view of the experimental lay-up cell components. The items shown in dashed lines are not part of the present cell but would clearly be present in a production version of the cell. Operation of the cell is controlled and monitored by a supervisory computer under the direction of an operator. The computer in use is a Cadmus 9000 series multi-user microcomputer based on the 68010 microprocessor with a Unix operating system and the C programming language. It communicates directly with all of the devices in the cell, co-ordinating and sequencing their operation, receiving information from sensors and sending instructions to the vision system, vacuum gripper and robot control computer. In the experiments carried out to date, this computer has also performed the necessary co-ordinate transformation calculations on data from various systems.

A network link is available for communications between devices in the cell and this would be the recommended data path for a production cell. However, for the present work, only direct device links have been used. This arose because with the robot installed at Short Bros, and much of the remaining equipment normally resident at the University, a permanent network installation at the robot would not be justified for research use alone.

Off-line programming of the cell robot with component data from the CAD system is presently achieved using the technique described in a previous paper (3). However, work is also underway using the robot simulation language GRASP (9) to provide an alternative off-line programming procedure. This particular package offers a sophisticated environment for robot program development and testing before transfer to the actual robot controller.

Sensory control of the vacuum gripper is achieved using a vision system as the source of sensor data. The vision system is an Autoview Viking system from British Robotic Systems. This is a stand alone vision development system based on a DEC LSI11/23 computer with a framestore board (to hold 256 x 256 pixels with 256 grey levels) and provision for up to four camera inputs. It is equipped with a wide range of vision processing algorithms to perform elementary operations on the captured data. Macros may be written which select and sequence a range of these basic operations to extract useful information from the picture data. The software, as provided, only processes images up to 128 x 128 pixels and is disc based which greatly extends the processing times. However, in spite of these limitations, the wide range of processing software provides an excellent base for experimenting with the analysis of image data. Two cameras are shown attached to the vision system in Fig. 6. This was the configuration used in the preliminary tests described later.

One difficulty with using the vision system in the cell structure shown in Fig. 6 was the lack of provision for external computer communications and control. As a stand alone development system, the only access provided was via the operator interactive terminal. Rather than attempt to redesign the hardware interface, which would have required a significant revision of the operating software, this problem was resolved by redirecting the terminal communications to the supervisory computer and writing appropriate device handling software to extract the required information. This problem of computer to computer communications was not unique to the vision system but proved to be a general problem. As a result, much time was spent in devising both hardware and software solutions. Of course such communications problems exist throughout all computer based manufacturing systems. It is to be hoped that the recent developments in standards for communications such as MAP (10) will quickly lead to greater compatibility between devices.

Fig. 6 makes reference to other sensor systems besides the vision system. In all the experiments performed to date, only vision system data has been used for control of the vacuum gripper. This provides two dimensional image data which has proved adequate for the work carried out so far. It can also offer an estimate of the third dimension (used as a rangefinder) but for more precise measurement in this direction, either another camera view or some alternative sensor system is required. One system which has been developed for the drilling project of the research programme uses 3 LVDT displacement transducers to sense both orientation and distance from a surface (1). This system is likely to be useful also for lay-up directly into the component mould.

4 ONLINE CONTROL

The sensory information derived from the vision system on errors in positioning of the gripper and/or the prepreg ply is used to modify the stored robot program in a way which allows the robot to reduce the errors. This results in a form of closed loop control of the gripper. However, unlike conventional closed loop control, this corrective action is only applied to particular points in the robot work cycle, when in close proximity to other devices. In the case of a lay-up cell it is used during the lay-up of a ply in a stack or in the component mould. At all other points in the cycle, the inaccuracies of the off-line program and the robot mechanism are normally small enough to provide acceptable path trajectories without danger of collision. We have used the term ONLINE CONTROL for this intermittent form of closed loop control of the robot end effector.

The implementation of online control is performed by the robot control computer. The sensory data on positional errors is transformed into required changes in robot co-ordinates by the supervisory computer and supplied to the robot control computer as a short relocatable subroutine sequence comprising only two program points. This is transferred to the T^3 robot using the External Function provided with the robot control software and the sequence executed. A further picture of the revised gripper position is captured and the complete process repeated until either satisfactory positioning is achieved or the requested movements of the robot are smaller than the available resolution will allow.

5 AUTOMATIC STACKING OF PLIES

Some preliminary tests have been performed to investigate the precision with which plies can be stacked one on top of another in the presence of positional errors. The plies are picked up from a flat surface, nominally the cutting surface and stacked on a flat surface, the lay-up surface. Fig. 7a shows the ply pick up locations as described to the off-line programming procedure whilst Fig. 7b shows the actual ply positions used which include significant positional errors. This test was designed to examine the online control and sensory signal processing performance rather than to emulate pick up from a cutting table. In practice, the prepreg cutting operation is unlikely to introduce any disorder of the plies. The actual lay-up position for the plies as indicated by the position of a 25mm diameter marker disc on the lay-up surface was offset about 20mm in both X and Y robot co-ordinate directions from the offline program description to cater for the practical case in which a lay-up mould is unlikely to be repositioned exactly between lay-ups.

The overall operational sequence of the task is as follows:-

Step 1 The robot moves to the cutting table and picks up the first ply using the program produced off-line. The relatively course resolution of the gripper allows it to cope with the degree of disorder deliberately introduced.

Step 2 The underside of the gripper is presented to the first camera positioned en route to the lay-up area and a picture frame is captured for locating the position and orientation of the ply on the gripper surface. This is determined by the vision system with respect to four reflective discs attached to the corners of the gripper at known separation distances. This information is passed to the supervisory computer.

Step 3 The robot moves to the off-line program lay-up location and pauses above the lay-up surface at a height which is sufficient to ensure that the lay-up marker disc is within the field of view of a second camera mounted on the robot arm and adjacent to the gripper. Analysis of the image from this camera locates the centroid of the marker disc and its displacement from the camera axis which identifies the desired alignment position. This information is used to correct the robot position.

Step 4 The robot is moved closer to the lay-up surface so that the image of the marker disc will almost fill the picture frame and another picture taken. Vision analysis of this picture determines any further robot corrections required. This step can be repeated if necessary until the remaining errors are smaller than the robot is able to resolve. For the hydraulic T^3 used, this resolution is about 0.5mm.

Step 5 The robot makes a final displacement correction for the position of the ply on the gripper and deposits the ply.

Step 6 The previous steps are repeated for the second ply.

In a preliminary test series, this lay-up sequence was repeated for a total of 10 trials and the magnitudes of the relative errors between the two plies were measured as shown in Fig. 8. Clearly the large, deliberately imposed errors have been mostly removed although there is still some residual error. This remaining error appears to arise from at least two sources:-

1. The vision analysis performed in step 2 introduces uncertainty about the exact location of the ply on the gripper surface as a result of poor resolution of the captured image. With the calibration marker discs located at the corners of the gripper, it is necessary to view the complete gripper surface. The square gripper surface and rectangular picture frame produces a useable window of only 128 x 90 pixels. This yields a picture resolution of 4.16mm x 6.26mm per pixel which is very course. We estimated the uncertainty in locating the centroid of the ply expressed in robot co-ordinates to be +/- 0.52mm in the X direction and +/- 0.32mm in the Y direction using an elementary statistical analysis for the given ply size. However, a review of the vision system software confirmed that, although our estimate was reasonable, the software actually returns the co-ordinates of the centroid in whole pixel units. As a result, the data supplied to the supervisory computer has an uncertainty range of +/-2.08mm in the X direction and +/-3.13mm in the Y direction. Clearly, this source of error can be reduced by improving the resolution of the captured image. This may be achieved by increasing the resolution of the vision system to 256 x 256 pixels or even greater. Doubling the number of pixels along each axis will double the resolution and reduce the measurement uncertainty accordingly. However, there would be a considerable increase in processing time which may prove unacceptable unless the analysis were performed in hardware. Alternatively, the ply location calculations could be performed in the supervisory computer where real number arithmetic would restore our estimated uncertainty. However, this has not been done since any production version of the cell would replace the vision development system with one specified for the task, thus avoiding the problem. Apart from the vision system, it would seem useful to introduce a matrix of calibration marks on the gripper surface rather than just at the corners so that the picture frame can be filled with the ply image whilst retaining reference marks in the field of view. This would be particularly effective when lifting small ply sizes.

2. The accuracy of the hydraulic T^3 robot mechanism was demonstrated in separate tests to vary quite significantly throughout its workspace. Numerous tests were performed to observe the response of the robot to requested displacements along each of the major axes. These showed an erratic behaviour with errors ranging from 0 to around 1mm, and with no simple pattern to the error profile. Clearly, for more accurate work, a better robot drive mechanism would be required.

6 CONCLUSIONS

The facilities described here form the basis for a comprehensive examination of the use of robotics in the composite prepreg lay-up process. It would be unwise at this point to predict that fully automatic lay-up will follow quickly. Even our limited experience of the production of composite components has confirmed that not all component geometries will be suited to robotic manipulation within the lay-up mould. In this regard, the manipulative skills of a human will be hard to match. However, much of the activity within a lay-up cell can be automated and it is these aspects of the process that should show an early and useful return for research effort.

The hydraulic drive robot used in this work is not very accurate, nor is it representative of the performance that can be achieved by modern electric drive systems. However, if the advantages of off-line programming are to be fully realised, there is clearly a requirement for good kinematic design and control of robot mechanisms.

Finally, the work has made clear to us the sad state that exists in the area of communications between computer-based manufacturing equipment. It seems imperative that a communications standard such as MAP is accepted and implemented quickly so that industry at large can gain from easy access to more integrated manufacturing systems.

ACKNOWLEDGEMENTS

The authors gratefully acknowledge the support of the ACME Directorate of the Science and Engineering Council and Short Bros plc, in this and other research projects into the application of robots in composite component manufacture.

REFERENCES

(1) Francey, S.D., Armstrong, P.J. Sensor-based robotic drilling for the aerospace industry. UK Robotics Research 1984, I.Mech.E., London, Dec. 1984, pp85-91.

(2) Campbell, N.A., Reid, I.M., McClean, J.H. Shorts robotic ultrasonic scanning system. UK Robotics Research 1984, I.Mech.E., London, Dec. 1984, pp71-77.

(3) Curran, J.P., Wright, E.J. Off-line programming and control of an industrial robot using a microcomputer. UK Robotics Research 1984, I.Mech.E., London, Dec. 1984, pp79-83.

(4) Allison, H.B., Pitman, W.A. Future trends
     in Technology and Automation.
     AIAA International Meeting, May 1980,
     Paper No. 80-0922.

(5) Wehrenberg, R.H. Automated Manufacturing -
     the future in aerospace Composites.
     Mechanical Engineering, Jan, 1980, pp46-50.

(6) Rapson, R.L. Advanced Composites - Evol-
     ution of Manufacturing Technology.
     AFWAL Material Laboratory, Wright-Patterson
     AFB, USA.

(7) Bettner, T.J. Fabrication of Aircraft
     Components using preplied broadgoods layed-
     up in the flat and subsequently formed;

     Cost benefits and Resource Utilization
     Enhancements.
     14th National SAMPE Technical Conf.,
     Oct. 1982.

(8) Marsh, A.K. McDonnell Douglas cuts Aircraft
     construction time.
     Aviation Weekly and Space Technology,
     Jan 9 1984, pp71.

(9) GRASP Reference Manual (2.00)
     BYG Systems Ltd., Highfields Science Park,
     Nottingham.

(10) Garrett, A. Manufacturing leads the trend
     in adopting OSI.
     Computing, Nov. 14 1985, pp28.

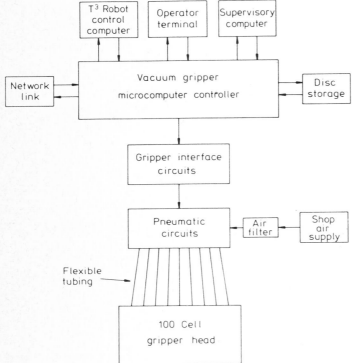

Fig 1   Functional elements of adaptive vacuum gripper

Fig 2   Vacuum gripper mounted on Cincinatti T³ robot

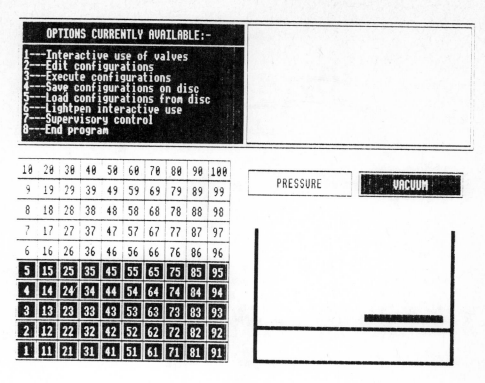

Fig 3   Screen display of vacuum gripper controller

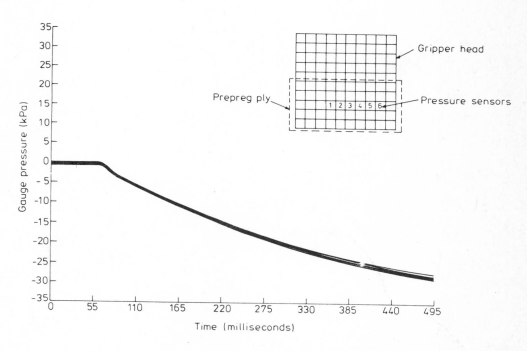

Fig 4   Gripper head pressures with application of vacuum

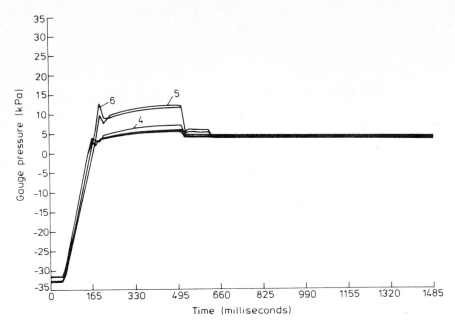

Fig 5   Gripper head pressures with change from vacuum to
         positive pressure supply

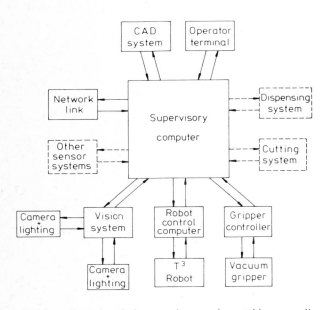

Fig 6   Functional elements in experimental lay-up cell

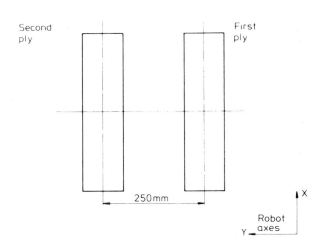

Fig 7a   Off-line programmed positions for ply stacking test

Fig 7b   Actual ply positions used in stacking test

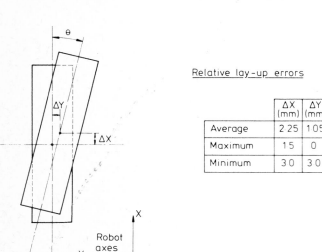

Relative lay-up errors

	ΔX (mm)	ΔY (mm)	θ°
Average	2.25	1.05	0.8
Maximum	1.5	0	0
Minimum	3.0	3.0	1.5

Fig 8   Results from ten trials of ply stacking test

46

Advances in gripper technology for apparel manufacturing

M SARHADI, P R NICHOLSON and J E SIMMONS, BSc, PhD
School of Engineering and Applied Science, University of Durham

Problems in the automation of apparel manufacturing are discussed. Gripper design for separation of fabric plies is identified as one of the important factors in the automation process. A study of existing gripper mechanisms is presented. A new gripper design is described together with preliminary test results.

1.   INTRODUCTION

In the manufacture of mass-produced apparel and the like, fabrics of similar types and shapes are often cut and produced in stacks. In production the individual plies of fabric are suitably joined in combination with other workpieces from different stacks to form sub-assemblies or complete garments. In such processing it is usually necessary to remove one ply from the stack and to deliver it to a desired assembly point. The material of each workpiece may be in one layer or in multiple layers.

The steps involved in automatically separating and removing the individual fabric workpiece are very difficult to accomplish reliably, in practice. Typical problems are for more than a single ply to be removed from the stack at one time, or during the removal operation for the rest of the stack to become displaced or disturbed in such a way that it cannot be properly or efficiently handled later. These difficulties arise because fabric plies tend to cling together in a somewhat unpredictable fashion. In particular, as a result of simultaneously saw or knife cutting the stack from laid-up fabric, fibres tend to entangle and form a mechanical interlocking bond or a strong friction bond, both of which make it difficult to separate and remove only the top ply. Also, with circular knit tubular fabrics fibre entanglement forms a strong interlocking bond between pairs of plies in a stack. The removal process is further complicated along the cut edge of a stack where the fibrous ends of the threads become entangled with the plies above and below. This engagement along the edges is particularly strong in an unprocessed cut stack because the cutting knife forcibly interlocks the cut fibres. Moreover, in the case of synthetic fabrics the thermoplastic yarns may be heat fused together when the pieces are cut from a stack.

One example of mass-produced batch organised apparel manufacturing is the assembly of standard mens briefs. It is simple in design yet involves many of the basic operations required to form more complicated garments. Also, its design does not change rapidly with fashion. In these respects it may be a good candidate for automation. However, the labour content of the finished product is less than 20% of the total cost. An automated system must be able to compete with the low labour content of that cost to justify financial investment. It must also save on material wastage of conventional assembly and improve product quality.

Fig.1 illustrates a twelve step breakdown in the assembly operations involved in manufacturing a pair of mens' briefs [1]. These include the following fundamental operations which demand practical solutions in order for a fully automated manufacturing system to be successful.

1. Cutting,
2. Destacking,
3. Limp material handling,
4. Sewing operations on straight, angled and contoured seams,
5. Joining operations on side-by-side and overlaid workpieces,
6. Multi-dimensional sewing operations,
7. Application of binding and elastic materials,
8. Application of motifs and labels,
9. Constant supervision of assembly operations,
10. Packaging.

At this stage of development, it would be an immense task to try to automate the entire manufacturing process. The experience of other research workers has shown that such attempts are not likely to be very successful [2]. A better approach to this problem would be to identify the critical and labour-intensive components of the work and then design automated sub-assembly systems to perform these tasks. Full automation would only succeed if the constituent components are proven to operate efficiently and reliably in an industrial context.

With modern engineering techniques it is possible to achieve a certain degree of automation in this area. Automation of cutting and packaging have been very successful and are currently used by many manufacturers. Semi-automated mechanical work aids such as control of the direction of sewing, positioning of workpieces, etc, have saved handling times and reduced labour costs. Also, microprocessor based sub-assembly techniques, such as elastic feed systems, have improved garment quality and reduced operation times.

Any serious attempt at full automation of batch organised apparel manufacturing would have to offer a credible practical solution to a major obstacle, which is the problem of ply separation and fabric manipulation. It is, of course, possible to avoid the problem of ply removal by re-organising the entire assembly operation on single ply processing [3]. However, because of heavy investments in the traditional assembly operations batch organized manufacturing is likely to remain the dominant process for the foreseeable future.

2. FACTORS INFLUENCING GRIPPER DESIGN

It is frequently possible to adjust and adapt one or more of the existing gripper designs to a reasonable level of efficiency for operations in which all conditions are fixed, with a single size and shape of ply and with the fabric being at all times of the same material. However, in a more typical operation the equipment may be called upon to handle workpieces in a variety of sizes and shapes. In mens' briefs assembly, for example, the gripper is, in addition, sometimes required to remove a double layer of gussets for processing. Conventional designs are not suited for such operations.Even where operation involves only a single size and shape of fabric, like in shirt collar assembly, serious problems may be experienced in dealing with multiple fabric materials. Therefore, a gripper must be able to handle a variety of sizes, shapes, fabric materials and multiple-layer workpieces,if it is to have any significant impact on the automation process. It must also include means of transporting the successfully separated workpieces from the pick-up point to a placing station in a controlled manner and without disturbing them, in the process.

Speed of operation, simplicity of design, reliability and efficiency of the separation process are factors of great practical significance if the gripper is to compete with human operators.

3. EXAMINATION OF EXISTING GRIPPER DESIGNS

Previous attempts at gripper design have included a variety of means and techniques. For the purposes of this paper, they may be specified in four different groups as follows:

(1) use of air suction or adhesive tapes,[4].

(2) engageable needle - like elements alone or in combination with air pressure and/or suction, [5,6].

(3) producing waves on the uppermost ply of the stack by mechanical or air jet means, [7,8,9].

(4) stretching or tightening of the uppermost ply by mechanical means, [10].

The use of air suction for de-stacking does not yield satisfactory results for limp, porous materials and tends to disturb the rest of the stack. However, this technique can be applied in the transfer and manipulation of the removed workpiece. With air suction and adhesive tapes it is neither possible to separate two or more plies at the same time (if so desired), nor is there any certainty about the number of plies removed [4]. Frequently, the separation is not successful and the normal automatic removal is impeded, thus reducing the efficiency of operation.

The techniques employed in the second category include mechanisms which make use of metal brushes or cards. These have curved teeth attached to pressure plates or pressure wheels, with the teeth locally pressed into the uppermost ply to facilitate separation [5]. Mechanisms which rely on the exact length of needle penetration to achieve ply removal suffer from poor reliability, since the thickness of porous fabrics is very uneven. Included in this group of gripper designs are devices which apply a number of open claw-shaped grips uponthe surface of the top ply which then close with the fabric in between. An air jet is then directed through a tubular needle under the engaged ply, thus creating a pocket of air under pressure between the uppermost ply in the stack and the next piece [6]. In this case similar disadvantages are encountered as for the air suction method, since the penetraction length of the hollow needle has to be in exact proportion to the fabric thickness. Furthermore, introduction of air pockets and subsequent removal of the workpiece can disturb the rest of the stack.

The Cluett gripper device is a good example of producing localised waves on the uppermost ply of the stack [7]. The complete set-up is rather complicated but the pick-up device essentially consists of a rotatable, partially teethed gripping wheel which engages the top ply under controlled pressure. In the proximity of the wheel there exists a guiding device in the form of a slotted nipping shoe which presses on the stack, again under controlled pressure. Rotation of the wheel buckles the top ply upwards between the wheel and the shoe, thus creating a wave, the leading edge of which is drawn into the slot in the nipping shoe. Continued rotation of the wheel accumulates enough material in the shoe slot to facilitate a localised separation. Complete ply removal is accomplished by insertion of a plate in the already detached region between the uppermost ply and the rest of the stack. Disadvantages associated with this technique are threefold. Firstly, the initial separation is only achieved if a ply edge or, better still, a ply corner is engaged. Secondly, it is doubtful if such a mechanism could efficiently remove a set multiple of fabric plies. Thirdly, the mechanism has occasionally been known to damage the contact area.

More recently, researchers at Hull University have developed and successfully tested a special sensory gripper device using the principle of wave generation with an air jet, rather than contact by mechanical means [8,9]. An air jet is used to vibrate the uppermost ply of the stack whilst a fork-like gripper is inserted between the topmost ply and the underlying stack. An infra-red crossfire sensor detects when the topmost ply has "flipped" over the bottom jaw of the gripper. Once the separation has been achieved, the gripper is moved under the top layer which is then gripped securely along one edge and peeled back from the rest of the stack. Different air pressure settings are required to separate different types of material.

The main advantage of this technique is that the process is non-projecting, since there is no sharp mechanical penetration of the fabric. However, it is inherently slow, with an initial separation time of 2-3 seconds. For single ply separation it is reported to have a success rate (reliability) of over 99%. As with the previous technique, it is not clear how the gripper would perform for a set multi-layer removal.

The last type of known technique in gripper design relies upon mechanically engaging the uppermost layer then tightening or stretching the workpiece, thus detaching the desired number of plies from the stack. One example of this design comprises application of sharp projections along two opposite edges, which are kept under pressure, [10]. The projections are then moved apart laterally a sufficient distance in order to stretch the workpiece. Finally, the gripper is lifted, removing the engaged and tightened workpiece.

Among the gripper mechanisms discussed in this paper, the last technique is perhaps the most promising in many ways. It is certainly the fastest method of ply separation. Since it does not rely solely on needle projection to effect separation, a very efficient and reliable performance can be obtained. A further advantage of stretching action is that the length of needle penetration is not a critical factor. As long as this length is less than the thickness of the workpiece, a successful separation is possible. With this mechanism it should also be possible to separate reliably a set number of fabric plies in a single operation.

There are two main disadvantages associated with the simple technique just described. Firstly, the device relies on the stretching action alone to produce complete separation. In fact, this may not always be the case. When the fibre entanglement between adjacent plies forms very strong bonds or when there is significant edge fusion, complete ply detachment may not always be achieved. Secondly, the use of two sets of needle projections at opposing edges of the ply is not desirable. Reduction of this mechanical protrusion to one set, or complete elimination of the needle penetration would greatly enhance the performance of this technique. Also, the addition of sensory devices to detect incorrect separation would further improve its reliability.

4. STRETCH AND ROLL GRIPPER

In the past few years one of the main activities of the Automated Apparel Manufacturing group at Durham University has been the design and development of a new gripper mechanism [11,12,13]. The device has been successfully tested for mens' briefs sub-assembly operations, where low equipment costs, speed of operation and reliability are of great commercial significance.

The Durham gripper has benefited from an extensive study of the existing designs for automated ply removal and manipulation. It sets out to overcome many of the limitations inherent in these designs and offers performance characteristics which are particularly suited for applications in low-cost garment manufacturing.

The main reasons for choosing this particular design are as follows:

(1) Initial separation by ply stretching is perhaps the fastest technique. However, this method alone, would not achieve a very reliable operation.

(2) Ply stretching together with rolling up of the partially separated workpiece ensures a total breaking of the bonds between the uppermost ply and the rest of the stack, resulting in a very reliable operation. Further advantages of the rolling technique are that the rest of the stack is not distorted during the ply removal process and the workpiece itself is not disturbed, or some parts of it accidentally folded over, during the transportation or placing operations.

(3) The length of the material to be stretched may be adjusted to suit different types of fabric.

(4) The projection of the oblique pins is not a critical parameter. However, it may be adjusted to effect multi-layer ply removal.

(5) The reliability of separation is mainly independent of the size and shape of the workpiece, giving the device a wide application within the apparel industry.

The new device is based on the principle of ply stretching to effect initial separation, rolling to complete the removal process and to facilitate a rolled-up delivery method for workpiece transportation. The design described in this paper uses only one set of engageable oblique pins and employs a number of pressure sensitive and opto-electric transducers to ensure reliable operation and monitor the de-stacking performance.

The pick-up mechanism is best described by a series of illustrations as shown in Fig 2. The roller head comprises a cylinder of approximately 100 mm in length and 30 mm in diameter. A small area of the cylinder's surface accommodates a set of oblique pins, together with the receiver component, RX, of an infra-red photo-sensitive diode. The transmitter element, TX, an infra-red light emitting diode (LED), is fixed by means of a rod to the base of this roller head and is set a small distance away from the device. The roller head is capable of independent rotational, independent translational, and simultaneous rotational and translational motion, the movements being achieved by means of digitally controlled electric motors.

A feed table may be used to hold the fabric stack immediately under the pick-up head, as illustrated in Fig 2(A). The gripping surface of the feed table is normally set at an angle and accommodates an adjustable pressure transducer in a marked area where the top end of the fabric stack lies. The feed table moves vertically, also by means of a digitally controlled electric motor. Above the stack towards its lower end, a pressure pad is provided to keep the fabric under controlled pressure during the initial separation mode. The horizontal position of the pad is adjustable to facilitate workpiece removal from stacks of different materials. The feed table, in addition, carries on its surface a retro-reflective opto-electric switch to detect the presence of the fabric load.

After a fabric stack has been placed on the table it advances vertically until the uppermost ply makes contact with the pad and the roller head, as shown in Fig 2(B). Once the pre-determined pressures at the two control areas are established, further movement of the table is halted.

The gripper is now actuated stretching the uppermost workpiece and thus breaking almost all the fibre entanglements to the left of the pad, as illustrated in Fig 2(C). The ply depth sensor on the gripper examines the number of plies separated and makes a decision on the next action. In the case of incorrect separation, the next action is de-actuation, followed by a further two attempts at removal. Correct separation is followed by rotational as well as translational movement of the gripper which rolls up the detached workpiece, breaking all the remaining bonds, 2(D). When the gripper approaches the pressure pad area, the pad is withdrawn and rotational movement of the gripper is stopped, leaving a small length of the ply unrolled as shown in Fig 2(E). The slope of the feed table prevents any disturbance to the rest of the stack during this peeling-off process.

The placing station comprises an alignment table, an area of which consists of a set of small holes and a series of reflex opto-electric switches. This area is normally positioned over a vacuum pad, as shown in Fig 3(A). The detailed design of the alignment table is beyond the scope of this paper, but the table enables adjustment in x, y, and rotational axes to achieve absolute mechanical registration for onward processing. At the placing end, the reflex opto-electronic switches detect when the loose end of the removed workpiece passes over them. The vacuum is applied to grip the loose end of the ply as shown in Fig 3(B). The rotational movement of the roller head, this time, is switched on in the opposite direction. This has the effect of unrolling the removed workpiece over the

alignment table in a predetermined position without causing any disturbance to the fabric, Fig 3(D). Once the workpiece is delivered, rotational movement is halted, the roller head is transported to the pick-up point, and the feed table is advanced for the next cycle to begin, with the pins positioned as illustrated in Fig 2(A).

The settings of the pressure transducers on the feed table and the pressure pad, the angle through which the roller head rotates when the fabric is stretched, the distance between the roller head contact point and the pad, and the length of pin penetration are factors determined empirically for each fabric type.

5. OPERATIONAL RESULTS

The gripper mechanism has been exhaustively tested on a variety of sizes and shapes of fabrics provided by the Lyle & Scott factory at Gateshead. The types of fabric supplied were in stacks of four dozen plies of one and one rib tubular circular knit cotton and woven cotton material. The device was set up to separate only single plies of fabric - and 5,000 pieces were processed in all; 4000 pieces were gusset parts and the rest front panels. The reliability of separation was found to be better than 99%, within which 2% of the operation required further attempts at removal. The separation time defined for this mechanism as the period during which the fabric is stretched, which effectively determines if the ply is to be removed or if a further attempt is to be made, is approximately 0.5 seconds.

The gripper is now being tested for a variety of fabric types and multi-layer fabric removal. Results of these experiments will form the subject of another publication. The main features of Durham gripper are speed, reliability of operation, and its potential for low-cost implementation together with a wide application area. There is very little practical data available for most of the techniques discussed in this paper. The Durham device has the same reliability as that of the Hull mechanism. However, it is cxapable of a much faster de-stacking operation, 0.5s compared with Hull's 2-3 s. Both the Durham device and that suggested by Bijttebier [10] rely on sets of pins engaging the fabric. However, the Durham device is simpler, having only one set of these pins. In the great majority of cases the pin projections have no adverse effect on the fabric. Nevertheless, in processing some fine fabrics pin engagement may not be acceptable at all, in which case the Hull device would be a better alternative, since it uses a non-projecting technique. Although the Durham mechanism has not been fully tested for multi-layer ply removal or a wide range of size and shapes of stacks, it is potentially well suited for a wide range of applications.

6. CONCLUSIONS

Some of the critical elements involved in the automation of batch organised apparel manufacturing have been considered. In particular, the problems associated with ply removal from a stack have been discussed.

Conventional gripper designs and some new approaches to solving the problems have been closely examined. The principles of the new stretch and roll gripper device have been explained in detail, giving experimental results of the performance characteristics for processing a particular type of fabric material.

7. ACKNOWLEDGEMENTS

The authors are grateful for support received from Lyle and Scott Ltd, Gateshead, and the Science and Engineering Research Council in carrying out the work described in this paper.

8. REFERENCES

1 – Sarhadi M, Nicholson P.R, Brown J.C, "Automated Garment Manufacturing" Proc.of the 4th Polytechnics Symposium on Manufacturing Engineering, 23-24 May 1984, Birmingham, pp 34-43.

2 – Kurt Salmon Associates Management Consultants, "A Report on Technology for Apparel Manufacture to the EEC Commission", Dec 1979.

3 – Rennel R.W, and Haltof B, "Robotics in Garment Manufacture" Shirley Institute and Salford University Industrial Centre Ltd., November 1983.

4 – William R. Conner Sr. "Automatic Feeder for Workpieces of Fabric or the Like" US Patent Number 3,670,674, June 20 1972.

5 – Littlewood K.J, "Method and Apparatus for Handling Fabric Workpieces" U.K. Patent Number 1,533,536, November 29 1978.

6 – Oldroyd D, "Devices for Picking up Pieces of Sheet Material from a Stack of Pieces" U.K. Patent Number 1,218,433, Jan 6 1971.

7 – Morton K.O, "Nipping and Lifting Means for Fabric Sections and the Like" U.K. Patent Number 1,511,184, May 6 1975.

8 – Taylor P.M. et al "The Application of Robotics in Garment Manufacturing Industry", Proc. ACME Robotics Initiative, 3rd Grantees Conference, University of Surrey, 1984, Section 45.

9 – Taylor G.E, "An Adaptive Sensory Gripper for Fabric Handling", 4th IASTED Symposium on Robotics and Automation, Amsterdam 1984.

10 – Bijttebier, G.A.H, "A Process and Apparatus for Separating Supple Sheets from a Stack", U.K. Patent Number 1,443,498, July 21 1978.

11 – Sterling M.J.H, Sarhadi M, Nicholson P.R, "Automated Garment Assembly" Proc.ACME Robotics Initiative, 3rd Grantees Conference, University of Surrey, 1984, Section 41.

12 – Sterling M.J.H, Nicholson P.R, "Automated Garment Assembly" Proc. of the ACME Conference on Advanced Production Machines, including Robotics, Trevelyan College, Durham, 30th Sept-20th Oct 1985, p.30.

13 – "Fabric Handling", BTG Provisional Patent Application, Number 8,521,217 August 1985.

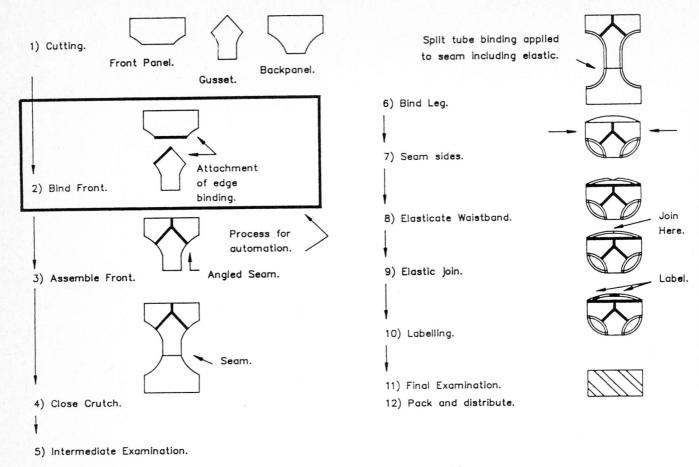

Fig 1   Complete Y-front garment assembly

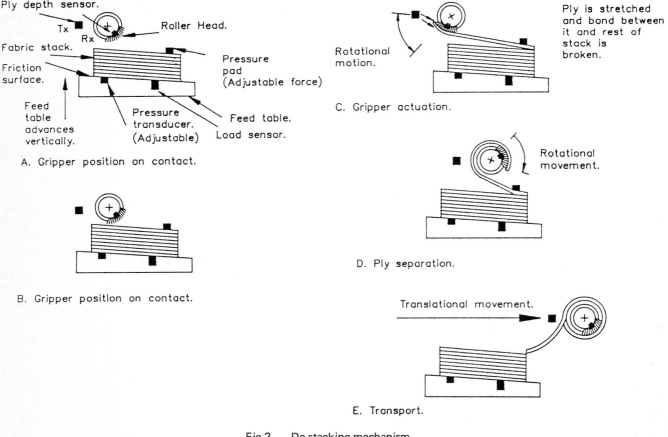

Fig 2   De-stacking mechanism

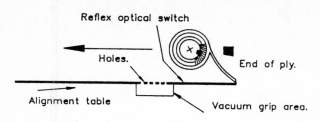

A. Gripper position at the placing end.

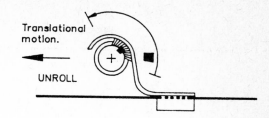

C. Ply release.

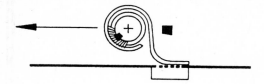

B. Vacuum is applied.

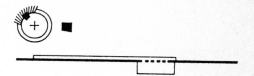

D. Workpiece delivered for processing.

Fig 3   Workpiece release stage

C368/86

An ultrasonic vision system for use on mobile robots and automated guided vehicles

P D NORMOYLE
Department of Mechanical Engineering, Trinity College, Dublin
J P HUISSOON, BA, BAI, PhD
Department of Mechanical Engineering, University of Waterloo, Ontario, Canada

1,1   INTRODUCTION

Automated guided vehicles have become recognised as integral components in fully automated manufacturing systems. Frequently their use is justified in more conventional manufacturing where transport of components or subassemblies between workstations or to and from storage areas is performed by standard forklifts.

The preferred means of providing guidance for AGVs is the inductive wire guidance technique whereby the vehicles travel along predetermined paths, these paths being permanently embedded in the surface over which the vehicles travel. Besides restricting the flexibility of enabling alterations to the guide path to be made, these systems are expensive to install (1980 estimate of £100 per metre) and frequently suffer wire breaks (54% of users [1]). Furthermore, the vehicles are constrained to travel only along these predetermined paths, any obstruction in the way necessitating the only available action : stopping. Autonomous free ranging vehicles could overcome many of these problems and provide real system flexibility.

The requirements for a true free ranging ability are that the vehicle control system be "aware" of its environment and that decisions based on this environmental information may be made. To enable information regarding the surroundings to be obtained, sensors capable of detecting both stationary and moving objects at distance of at least a few metres from the vehicle are required. Furthermore, acquisition and analysis of the sensor data in real time must be feasible if the vehicle is to move at a reasonable speed.

Two principles on which suitable sensors could be based are light and sound transmission. In general, optical sensor systems require considerable data processing capacity for the necessary analysis to be performed at sufficient speed. Such systems also tend to be expensive. Ultrasonics has the ability to overcome some of the limitations imposed by the video camera; object range is inherently available whereas in optical imaging it is not. Ultrasonic sensor systems maybe created that are considerably less expensive and yet provide sufficient information for basic guidance purposes, where more importance is attached to detecting the location of an obstacle with respect to the vehicle rather than providing detailed environmental information.

This paper describes an ultrasonic based imaging system that allows multiple objects in a 65 degree field of view to be detected and a "map" of this area to be generated.

1.2   ULTRASONIC RANGING

Several research papers have appeared describing air coupled ultrasonic imaging techniques for application in robot vision and blind aids [2]. Many of the methods involved suffer from over-sophistication or have other drawbacks which limit their effectiveness in locating obstacles.

In its simplest form ultrasonic ranging may be performed by emitting a short tone burst and determining the time taken for its echo to be received [3]. Since the speed of sound in air is relatively independent of most environmental variables (except for temperature which may easily be measured), the distance to the object may be determined if the "time of flight" is known. An alternative technique, also based on the echo delay time but using a continuously transmitted frequency modulated signal, may be used [4], but was found less suitable with the available transducers and posed analytical problems when multiple objects were encountered.

Phased array systems are common in medical imaging [5] [6], and have been developed for operation in air [7]. The main disadvantage of this method is that a large number of closely matched elements is required if narrow beam patterns are to be achieved.

In the present system the area immediately ahead of the vehicle is divided into twelve segments and the pulse echo technique used to measure target object range in each segment. The angular position is determined by comparing the time of flight as recorded by a pair of receivers positioned either side of the transmitter. The distance between the receivers being fixed enables a microprocessor to compute the objects coordinates using geometric relationships. In a sense this is similar to the stereo disparity effect that is used in optical systems.

2.1   SYSTEM CRITERIA

The main criteria considered in the prototype design may be detailed as follows :-

(1) UPDATE RATE : Range measurement update rates sufficiently fast so that information gathered does not become seriously out of date at

typical vehicle speeds of up to 1m/s. Since the velocity of sound in air is relatively slow (343 m/s @ 20 C), the rate of update will always be a trade-off with maximum range requirements.

(2) ANGULAR RESOLUTION : For multiple objects ahead of the vehicle to be perceived as separate targets narrow transmitted beam patterns are required. This is also desirable if narrow openings such as doorways are to be identified without the system becoming confused by echoes from the surrounding walls. Narrow beamwidths in the vertical plane reduce unwanted echoes from floor and ceiling.

(3) DETECTION PROBABILITY : The unit is required to detect the presence of a wide variety of target types and textures. Single frequency echo location is prone to partial echo cancellation due to the effects of destructive interference encountered with some types of target. Improved target detection probability can be obtained by transmitting more than one frequency [8]

(4) COST : This is important since there is little point in developing a sophisticated system which turns out to be too expensive to incorporate on a production vehicle. The prototype makes use of cheap readily available transducers.

3. SYSTEM DESIGN

3.1 MECHANICAL : Figure 1 shows a plan of the area in which objects may be detected. Any object is referenced in polar coordinates, namely range and angular ordinate with respect to the transmitter.

Since the returned echoes will be at the same frequency as the transmitted tone (neglecting the doppler effect due to any relative motion), multiple targets may result in an indecipherable series of received signals. Although this may be overcome using a frequency modulated transmission [4], the data processing required was deemed to be excessive for this application. In addition, covering the desired 65 degree field of view (which was considered a necessary minimum) is not possible using a single transmitter due to the relatively narrow beam pattern that is characteristic of available ultrasonic transducers. Using a number of transducers may suitably widen the beam but also results in interference patterns between signals. The solution adopted was to rotate a single transducer through a 65 degree arc, transmitting a tone burst every 5.4 degrees. This enabled a uniform coverage (in terms of SPL) of the required area which may easily be extended, if necessary, to give a wider field of view. It also minimises the number of targets that will return an echo for any one transmission and allows the microprocessor to identify false echoes or estimate the target position if only one sensor receives an echo. An encoder mounted on the rotating shaft of the transmitter, allows the angular rotation to be measured. The transmitter/ encoder combination is continuously driven, via a four bar mechanism, by a geared DC motor rotating at 1 revolution per second.

3.2 ELECTRICAL

Figure 2 shows the electrical schematic in block form. The encoder has two output channels in quadrature, each channel giving 100 cycles per revolution. By using both rising and falling edges of each channel signal, pulses may be generated at 0.9 degree intervals of revolution. Two limit switches are used to define the scan limits, and ensure correct initialisation on each scan. Every sixth pulse initiates a transmission cycle composed of 2 frequencies : 13 cycles at 51kHz followed by 15 cycles at 39.7kHz. The transmitting transducer used is an electrostatic device manufactured by Polaroid and quoted at 110dB at 1 metre. The full angle beamwidth at 50kHz is approximately 10 degrees at the -3dB radial. Beamwidths down to 2-4 degrees are achievable with the addition of commercially available acoustic horn/lens assemblies.

Signals from each wide angle receiver are amplified, band pass filtered and compared against a time dependent threshold. This threshold is desirable since unwanted echoes from close objects tend to be much stronger than wanted echoes from far targets. It also provides some immunity from strong echoes returned from outside the maximum range overlapping. Once the threshold is exceeded the comparator sets the SR flip flop whose output is used to gate clock pulses to the counter. The flip flop is automatically reset by the following transmission pulse. This mode of operation enables only the first received echo to be detected, multiple echoes arriving after the first are locked out.

As each burst is transmitted, the microprocessor starts a time out clock corresponding to the time interval for maximum range. At time out data from the two 16 bit counters is latched and read into memory whether an echo has been received or not. Other inputs to the processor include transmission pulse count and current scan direction. Time out and data read in routines are implemented in Z 80 assembly language. Echo timing routines could also be implemented in assembly language, but clock frequency is then limited by the processor speed.

3.3 DATA PROCESSING

At the end of each scan 12 sets of data have been stored corresponding to the ranges measured by the two receivers in each of the twelve segments. The time interval between scans is used for data processing. Each count stored corresponds to the total path length travelled by the ultrasonic pulse to each receiver. This is shown in Fig.3 where the object detected is assumed to be a point target at a distance R from the transmitter and an angle 0 from the centre line. The count for each receiver thus corresponds to the time taken for sound to travel (R+P1) and (R+P2) respectively, from which R and 0 may be calculated the distance d being known. In the event of only one receiver detecting an echo (due perhaps to target geometry), R may be calculated using the measured value of transmitter angle.

Due to side lobes in the radiated sound pattern, highly reflecting objects not directly in the main beam, may return an echo of sufficient strength to trigger the comparators. Since only the first echo is accepted by the system, this may mask another target directly in the main beam but further away. In the present system, the value of O computed is compared to specified limits on either side of the direction in which the transmitter is facing so that the possibility of the presence of an undetected object is not ignored. This also applies in the case where more than one object is present within the designated beamwidth but at different distances from the transmitter. This system may be improved upon by accepting all echoes that occur within a suitable time period after transmission and analysing all the permutations of possible target locations.

The target locations are computed using a combination of formulae and look-up tables. At present, these are expressed in Cartesian co-ordinates so that a visual map of the field of view may be displayed on a VDU. The same information may be used to control the motion of an AGV.

In cases where the calculated coordinates fall outside the preset limits of a given segment, the data is rejected and substituted with a new calculation based on the current angle of the transmitter. This compensates for the fact that an error of 9 degrees will occur if only the 40kHz component is returned to one receiver.

4. SYSTEM PERFORMANCE

Testing was carried out with the clock frequency set at 50kHz, giving a resolution in path difference measurement of 6.9mm. For the present receiver spacing of 60cm, path difference is approximately 10mm per degree offset from the centre-line. Two factors limit measurement accuracy :-

(i) Thermal fluctuations along the ray paths

(ii) Finite rise time of the receivers. This results in strong echoes being detected fractionally earlier than weak echoes

The accuracy in range measurement is limited by how well the temperature in the insonified space can be measured. A 10 degrees C change about a mean ambient temperature of 20 degrees would give approximately a 2% error in range measurement. Repeatable accuracy in the laboratory at 3 metres was +/- 1 cm. Clearly as the distance is reduced accuracy can be increased since the variation in propagation velocity is less important.

Trials with hard cylindrical targets placed at measured distances from the transmitter showed that angular position may be calculated to a precision of ± 1.5 degrees. This falls off to ± 4 degrees near the scan limits due primarily to the decrease in receiver sensitivity at wide angles. Results obtained with larger targets were found to depend very much on target type, orientation and surface topography. This is shown in Figure 5 where a target returns echoes to the receivers from two different points on its surface, resulting in an error in determining the range of the nearest point. This could be minimised at the expense of scan speed, by dividing the scan area into a larger number of smaller steps. Worst case positional errors (errors between actual and calculated coordinates) with a wide variety of targets were 25cm, considered adequate for obstacle avoidance applications. Detection probability was 100% for all targets closer than 2 metres.

Measurement accuracy is considerably improved by increasing the transmitter power, however, with the present setup this results in direct pickup at the receivers and necessitates a blanking period. Receivers with an improved risetime would also allow a higher accuracy or closer spacing between the transducers.

CONCLUSIONS

A stereo ultrasonic sensor system has been described that may be used to provide environmental information for the control of an automated autonomous vehicle (AAV). The current system enables an updated map of the area ahead of the veh. to be generated twice per second. The frequency at which this may be performed is limited by the maximum range at which objects are to be detected and the minimum separation between adjacent objects that needs to be resolved. At present, further work is underway to refine the ultrasonic ranging system for combination with a low cost vision sensor to improve the resolution and operating speed and thus provide improved control for guidance purposes.

REFERENCES

[1] Redmond, B. "Driverless Truck Progress": Materials Handling News, April 80.

[2] Heyes, T. "Sonic Pathfinder", Wireless World, April 84, 26-29

[3] Mouchoud et al. "A Self Adapting low cost Sonair for use on Mobile Robots" Sensor Review, Oct. 81, 180-183

[4] Kay, L. "Airborne Ultrasonic imaging of a Robot Work Space". Sensor Review Jan. 85, 8-11

[5] Yuen et al. "Microprocessor Systems and their application to Signal Processing", Academic Press 1982, 282-297

[6] Gregus, P. "Ultrasound Imaging", Focal 1980

[7] Kudora et al "Ultrasonic imaging System for Robots using an Electronic Scanning Method", Robotica 1984, Vol. 2, 47-53

[8] Polaroid Ultrasonic Ranging Kit Manual Polaroid Corp. 1984.

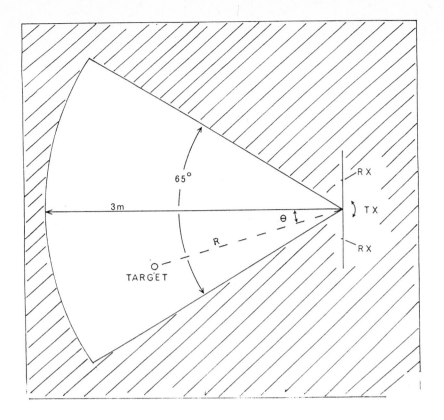

Fig 1   Transducer layout and scan area

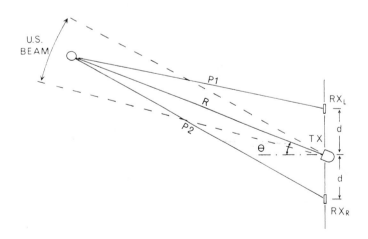

Fig 3   Target measurements

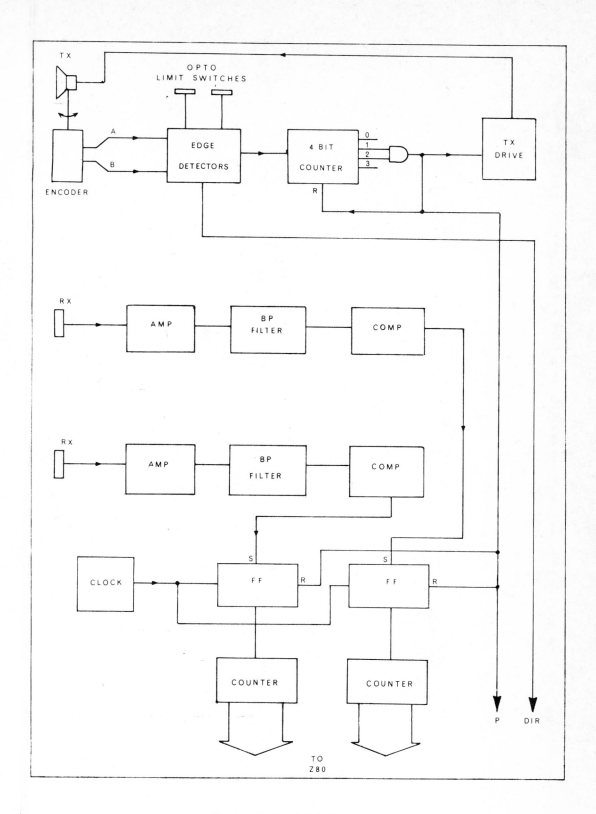

Fig 2   System block diagram

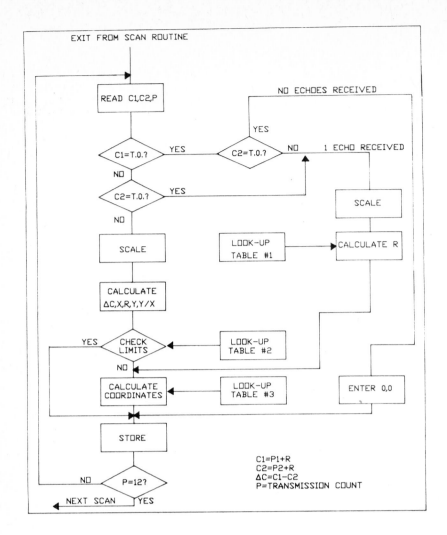

Fig 4   Data flow chart

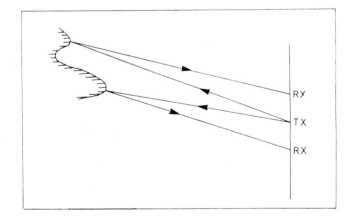

Fig 5   Detection of a complex target shape

Real-time collision-free path calculation for robots

N BALDING, BSc and **C PREECE**, BSc, PhD, MIEE
School of Engineering and Applied Science, Durham University

SYNOPSIS This paper presents a method for automatically generating collision free paths between obstacles for a revolute robot. The world model is represented as a collection of spheres, some overlapping, and the robot is represented by connected cylinders.

The aim of the method is to provide an on-line, real-time capability for collision free path calculation in a flexible manufacturing environment.

1 INTRODUCTION

The use of robots for pick and place, and assembly operations has advantages over dedicated handling machinery in that robots may be reprogrammed to carry out different tasks. This property is used in many areas of industry including Flexible Manufacturing Systems (FMS) where small batches of parts are produced.

However different tasks require recalculation of the robot paths. Currently robot paths may be programmed in one of three ways:

(a) Lead by the nose. The required path is tracked manually and the robot controller records the coordinates.
(b) Teach Pendent. The robot is moved over the required path by manual controls, again recording the coordinates at discrete intervals.
(c) Off-line programming. The path is defined using a computer simulation of the robot and its surrroundings. The coordinates are then loaded into the robot control computer.

Developing robot programs can be an expensive and tedious task. The time and money required to write a robot program must be less than that needed to carry out the task manually. When robots are used for repetitive jobs the cost of programming is spread over many operations.

This paper presents a new method by which robots may be automatically programmed. It has the advantage that the programming cost for new paths is eliminated so that the robot may be used in situations where the task is changing, with paths automatically reprogrammed between tasks.

A path planning computer contains a model of the robot and its surroundings. It also holds the current state of the robot and a model of the robot's articulation. On receipt of a task specification, the path planning computer calculates an efficient collision free path for the movement of the robot. This is transmitted to the robot control computer, which controls the robot itself.

The system has been implemented for test purposes, using an Intel 8086 based microcomputer together with a Smart Arms 6R 750 robot. In the present state of development, the time of calculation for simple path finding problems matches the robot execution time. Thus for a sequence of different tasks the new paths can be calculated while the previous path is being executed. This on-line operation of a path finding method is an important aim of the work.

2 APPROACH TO THE PROBLEM

2.1 Background

Automatic pathfinding for robots was investigated by Pieper (1) in 1968. Since then the major contributions have come from Udupa, Lozano-Perez, and Brookes (2,3,4). These authors all use polyhedral models to represent obstacles. This work has been extended most recently by Luh, Gouzenes, Chien and Gilbert (5,6,7,8).

An alternative approach has been reported by de Pennington et al (9) where the robot is modelled by a series of interconnected spheres. This is further detailed by Balila (10).

This paper presents a method of modelling obstacles by spheres, in an attempt to simplify the problem for practical applications.

2.2 Requirements

In most published work dealing with off-line path finding techniques, the time of calculation is not critical. The path information is down loaded to the robot controller for use in the manufacturing process.

The aim of this work is towards an adaptive scheduling system in which the full advantages of a truely flexible manufacturing system are realised. The scheduling program needs to be tolerant of a wide range of variables relating to the work in hand, and to have the ability to cope with changes in the work pattern.

Information from sensors on the plant may be used to influence the scheduling, as will the introduction of a new component design into the existing work pattern.

In a highly flexible system, such changes require dynamic re-scheduling which involve fast re-calculation of robot paths. The extent to which this can be achieved will determine the re-scheduling capabilities of the flexible manufacturing system.

In order to achieve real-time operation a compromise must be made between the efficiency of the calculated path and the calculation time. For any path planning problem there is an optimum solution based on a chosen cost function. Operational constraints may make a faster sub-optimal solution more acceptable in a particular application.

The method adopted here produces a sub-optimal path using a simplified world model. Any sub-optimal solution must ensure that the calculated path is a safe one, that it satisfies the "collision free" criteria, and that any divergence from the optimal path tends to produce greater rather than smaller clearances.

3 REPRESENTATION OF THE ROBOT AND ITS SURROUNDINGS – THE WORLD MODEL

3.1 Modelling the robot's surroundings

Most published computer models of robot surroundings are in the form of polyhedral obstacles. This geometry is chosen because most obstacles tend to have flat surfaces and straight edges. However these model forms can be difficult to deal with in path finding calculations. Figure 1 is a diagram of a situation, where a robot has to find paths avoiding two box shaped obstacles. In this case the robot and the obstacles are all polyhedral objects and so a polyhedral model would imply a high degree of accuracy. However, a considerable simplification in the path calculation algorithms can be achieved by representing obstacles as collections of spheres.

Each sphere is represented by four items of data comprising the centre coordinates and the radius. These models simplify clearance calculations, which are reduced to finding distances between spheres. Figure 2 shows a model based on spheres which might be used to approximate the real situation of figure 1.

It is contended that in many practical situations, objects may be modelled satisfactorily by only a few spheres, without compromising the useful workspace of the robot.

3.2 The Robot

The complexity and detail of the robot can affect the efficiency of the path finding algorithm. As the accuracy of the robot model is increased, the workspace for the path finding algorithm is enlarged, but the benefit that this brings may be outweighed by the extended time of calculation. Simple models, which occupy more space, may produce longer paths, but the robot will tend to move further away from the obstacles, increasing rather than reducing the safety factor.

The test robot used in this work has two major links, the upper arm and the forearm. The upper arm is attached at the base of the robot and has two degrees of freedom, it can rotate about a vertical axis and about a horizontal axis. The forearm is attached to the upper arm by the elbow joint which rotates about an axis perpendicular to the upper arm.

The upper arm and forearm are both long thin members which are most simply represented as line segments. To do this the minimum bounding cylinders for the links must be found. The diameters of the obstacle spheres are increased by the radii of the cylinders.

3.3 Developing a Model

The model of the surroundings must be "safe". This means that every obstacle that is a candidate for a collision must be completely enclosed in obstacle spheres in the model representation.

Approximations may be made in modelling obstacles in the robot space. If a few small obstacles in the robot work space are close together and the space is not critical to the path of the robot, it is acceptable to enclose them all in one sphere rather than to model them individually.

Obstacles which are significant in the calculation of the path may be modelled by a number of spheres. The number is determined by the available workspace in the vicinity. In some cases, the extra space taken up by reducing the number of spheres representing an obstacle is not found to be critical, while the saving in calculation time may be considerable.

4 PLANNING

Planning robot movements can be a complex operation, but the proposed model of the surroundings simplifies the task. The planning problem may be considered in three stages, following Udupa (2).

4.1 Stage 1. Feasibility

The positions of the robot are checked for feasibility at both the start (S) and goal (G). Positions which are out of the robot's workspace or which would cause collisions with obstacles are clearly unacceptable.

The planning of robot movements is determined by stages 2 and 3, which are called approach path planning and mid-phase planning respectively.

4.2 Stage 2. Approach path planning

Approach paths are paths which move from positions with good clearance from local obstacles to end positions such as S and G. A path which moves to S, from a position clear of obstacles Sm, may also be followed in reverse when moving from S, to Sm. The positions which are clear of obstacles are generally close to the end positions so that the approach paths are short.

The positions Sm and Gm are used as the start

62

and goal positions for the mid-phase planning.

Approach path planning requires special knowledge of the environment and details of the chucks and grippers of the machines. As an example, consider the manipulation of a component into a lathe. The part must be lined up with the central axis of the chuck and must move along this axis until it is at the required position in the lathe chuck jaws.

Approach paths may be considered to be related to machine geometry, and calculated for the specific machine configuration. This part of the path planning is therefore separated from the mid-phase path planning and treated differently. The user defines an approach path by giving the following information:

1.   The orientation of the part for the approach path.
2.   A vector defining the direction, length and position of the approach path.

This information is machine and part specific. Special programs are written to calculate the approach paths from the above information.

4.3 Stage 3. Mid-phase planning

Mid phase planning calculates the collision-free paths between the robot configurations Sm and Gm as defined by stage 2. First the problem of finding a path for the upper arm is considered, followed by the calculation of the forearm path.

4.3.1 Mid-phase planning for the upper arm

The upper arm is modelled as a line segment fixed at one end to the origin. The obstacles are represented as collections of spheres. The aim of the calculation is that the path must avoid the obstacle spheres and it should be efficient.

The upper arm has only two degrees of freedom, elevation and rotation. This means that the problem may be viewed as a two dimensional one even though the arm is moving in three dimensions. A projection of the upper arm from the origin gives a point on a background of circles. The problem is transformed into that shown in figure 3.

We assume that the aim of the mid-phase path planning is to calculate the shortest collision free path between endpoints. Factors such as the energy consumed by the robot, the wear, and the time of movement may modify the optimal cost, and may be incorporated later.

Given this assumption it may be seen that the shortest path from a startpoint Sm to a goalpoint Gm consists of straight lines between obstacles and circular arcs when traversing the circumferences of obstacles.

One method of finding the shortest path is to calculate all the possible paths of this type. If there are n circles and the path passes around all of them once, then the maximum number of paths is given by equation (1).

$$2^n n! \qquad (1)$$

Thus the maximum possible number of paths m is

$$m = 1 + \sum_{i=1}^{i=n} 2^i i! \binom{n}{i} \qquad (2)$$

where $\binom{n}{i}$ is the binomial coeficient i of n.

This grows large very quickly, for n=3, m=79; for n=4, m=633; and for n=5, m=6451. In practice the number of possible paths is generally less than this, because some paths between obstacles will be blocked by other obstacles, but even so an unacceptable number of options may remain.

4.3.2 A Heuristic Approach

To determine the shortest distance path length a heuristic method of graph searching is quicker than the method of calculating all possible paths. The following method is based on that of Hart. (11)

A graph is defined to consist of a set of elements, called nodes, and a set of paths between nodes called branches. Each branch between nodes has a cost associated with it.

The sub-graph G_n is the set of nodes accessible from a particular node n. The sub-graph is calculated by a successor operator T.

A path from a start node to a goal node is an ordered set of nodes $(n_1, n_2, \ldots, n_k)$ with each n_{i+1} a successor of n_i. Every path has a cost which is obtained by adding the individual costs of each branch, $C_{i, i+1}$, in the path. The optimum path from n_i to n_j is a path having the smallest cost over the set of all paths from n_i to n_j.

Starting with the start node S, the subgraph G_s is generated by the successor operator T. During the course of the algorithm, if the subgraph G_n of a node is generated, then the node is said to be expanded. The minimum cost of each node encountered is calculated, and a pointer to the predecessor of each node along that path is stored. The unexpanded node with the minimum cost is always expanded next in the algorithm. Finally, the algorithm is terminated when the goal node G is reached.

4.3.3 Application to spherical obstacles

A path around a sphere, as seen by an observer, may be in either a clockwise or an anticlockwise direction. Thus for every sphere there are two nodes in the graph.

In the two dimensional case the shortest path will consist of straight lines between circles and arcs around the circles. In the real situation this corresponds to the movement of the robot arm in planes between spheres. When the robot moves around a sphere it will follow the surface of the sphere until it reaches the plane of its next movement. This will be tangential to the surface of the sphere.

Two cost functions were considered. Firstly the cost of a branch was defined as the value of its length. The lowest cost from S to G becomes the path of the shortest length.

The second cost function was defined as follows. The cost of a branch between node 1 and node 2 is the extra distance the robot has to travel along that branch and from there directly to the goalpoint, compared with a branch which goes directly from node 1 to the goal point. To calculate this cost the following equation is used:

$$
\begin{aligned}
Cf = \quad & \text{distance along branch from node 1 to} \\
& \text{node 2} \\
+\ & \text{direct distance from node 2 to the} \\
& \text{goal point} \\
-\ & \text{direct distance from node 1 to the} \\
& \text{goal point.} \qquad (3)
\end{aligned}
$$

A comparison of the two methods is shown in figure 4. For ease of representation figures 4(a) and 4(b) show graphs of independent nodes, the double nodes of circles are not considered. Figure 4(a) shows that with the first cost function, all the branches were searched. In this example the method took ten steps. Figure 4(b) shows that the second Cf calculation caused the search to be completed after only five steps. The second cost function was preferred, as the time saved by computing fewer steps was greater than the extra time taken to calculate the more complicated cost function.

The successor operator T generated subgraphs of a node n as follows. Paths to all the other nodes were proposed. The node which represented the same circle as n but the opposite direction was rejected. Of the candidate paths those which left the circle with angles A of less than 180 degrees were rejected because the optimum path must be tangential (see figure 5). The remaining paths were then checked for intersections with other circles, those that were clear, formed the subgraph Gn.

4.3.4 Forearm path planning

Having fixed the trajectory of the upper arm the locus of the elbow is established, and thus the problem of finding a path for the forearm becomes a 2D problem instead of a 3D problem.

The forearm path is constrained by the locus of the elbow and the elbow angle. A graph of distance along the locus of the elbow may be plotted against elbow angle. Viewed from the locus of the elbow, a graph of the two dimensional working space of the forearm can be calculated showing the positions of obstacles. The two variables which define forearm position are now elbow locus and elbow angle. Obstacles in space can be transformed into these coordinates and appear in the new two dimensional space as shown in figure 6. Here a spherical object appears as a complex shape due to the transformation process. However it is clear that a collision free path can be defined as one whose coordinates do not invade the shaded area.

The planning strategy adopted here is to define a "straight" path from the starting point Sm of the elbow locus to the goal point Gm and test for interception with the object. If this interception is detected, then a new path is proposed which avoids the collision. The "straight" path in this case is a linear function in the transformed space. The procedure is repeated until the final goal point is reached.

4.3.5 Planning of the gripper and workpiece

The gripper was modelled as a series of spheres. While moving in mid-phase motion the gripper is aligned with the forearm. If the gripper is small compared with the forearm this avoids the need to calculate the extra degrees of freedom for the gripper itself. The position, and representation of the workpiece in the gripper is known from the approach path planning.

4.3.6 Avoiding obstacles of the forearm, gripper and workpiece

The path planning for the forearm, the gripper and the workpiece is done together. The paths of the forearm and the spheres representing the gripper and the workpiece are calculated, and the closest point of the forearm, gripper or workpiece to the first obstacle is calculated. A path which avoids the obstacle is proposed as follows. The vector between the closest point on the robot and the centre of first obstacle is calculated. This vector is then extended so that the robot and the obstacle are a set distance apart.

5 IMPLEMENTATION

5.1 Description of hardware

The path planning algorithms are implemented on an 8086 based machine, called the path planning computer. This is interfaced to the robot control computer through a serial link. Figure 7 shows the system hardware and data links between each component.

5.2 Description of software

A flowchart of the system software is shown in figure 8.

The world model, represented by the sphere data, is entered by a separate program which stores the data on disc. The world model also includes the configuration of the robot. This data is then read by the path planning program and, if necessary, updated as objects are moved by the robot.

To test the system, a task description was loaded into the path planning computer. For a typical pick and place or an assembly operation the task description consists of the specification of the start position and desired goal position of the part.

The goal feasibility program checks that the end positions are a) within the robots workspace and b) clear of obstacles. If this test fails the program reports the error and halts.

The approach path program calculates paths to S and G, from Sm and Gm using information about part size and type of approach path required.

The mid-phase planning program then calculates the collision free path for the robot as described above. The path planning computer's definition of the path is post-processed into a form which the robot control computer can accept. For the test robot the path was divided into small line sections so that the straight line paths defined by the path planning computer were followed closely.

The data transfer was done on-line. The path planning computer waited for the robot control computer to flag it for the next set of path data. The data was then transferred via the serial link and as soon as it was stored, the robot computer started the path movement.

5.3 System performance

Tests have been carried out on pick and place tasks with objects strategically placed in the workspace. Coordinates were generated randomly for start and goal positions. While these tests are by no means exhaustive, they have shown that the system can successfully find collision free paths between the sample obstacles.

The compromises made to simplify calculation, in order to provide on-line real-time performance, have been shown to be acceptable for the test environments so far encountered. It is likely that further refinement of the algorithm will be necessary as more constraints are placed on the workspace.

The technique indicates that automatic real-time collision free path planning can be successfully accomplished using small micro-based computers.

6 CONCLUSIONS

The use of spheres to model real obstacles makes it possible to calculate collision free paths in real-time using a small micro-computer. The use of this modelling technique implies a compromise in the accuracy of representation of the workspace model.

Except in very tightly constrained operating environments this compromise has been shown to be acceptable, indeed the sub-optimal paths calculated tend to move further away from obstacles, and to increase safety margins.

The use of two dimensional transformations in the planning algorithm has been found to be of value in the calculation of collision free paths. The present approach decouples forearm and upper arm planning. In more restricted workspace environments, an iterative approach may be required. This is the subject of further investigation.

7 ACKNOWLEDGEMENTS

The authors wish to acknowledge support for this work from the Science and Engineering Research Council and Delta C.A.E. Ltd.

8 REFERENCES

(1) PIEPER, D.L. The kinematics of manipulators under computer control. PhD Thesis, Stanford University, October 1968.

(2) UDUPA, S. Collision detection and avoidance in computer controlled manipulators. PhD Thesis, California Institute of Technology, 1977.

(3) LOZANO-PEREZ, T. Automatic planning of manipulator transfer movements. IEEE Transactions on Systems, Man and Cybernetics, volume SMC-11, No.10, October 1981, pages 681-698.

(4) BROOKS, R.A. Planning collision free motions for pick-and-place operations. The International Journal of Robotics Research, volume 2, number 4, Winter 1983, pages 19-44.

(5) LUH, J.Y.S., CAMPBELL, C.E. Minimum distance collision-free path planning for industrial robots with a prismatic joint. IEEE Transactions on Automatic Control, volume AC-29, number 8, August 1984, pages 675-680.

(6) GOUZENES, L. Strategies for solving collision-free trajectories problems for mobile manipulator robots. International Journal of Robotics Research, volume 3, number 4, pages 51-65.

(7) CHIEN, R.T., ZHANG, L., ZHANG, B. Planning collision-free paths for robotic arm among obstacles. IEEE Transactions on Pattern Analysis and Machine Intelligence, volume 6, number 1, pages 91-6.

(8) GILBERT, E.G., JOHNSON, D.W. Distance functions and their application to robot path planning in the presence of obstacles. IEEE Journal of Robotics and Automation, volume RA-1, number 1, March 1985, pages 21-30.

(9) de PENNINGTON, A.,BLOOR, S., BALILA, M. Geometric Modelling: A contribution towards intelligent robots. Proc. 13th Int. Symposium on Industrial Robots and Robots 7 1983 Chicago North Holland Publ. Co. pp7.35-54.

(10) BALILA, M.A. Robot path planning using geometric modelling systems. Ph.D. Thesis University of Leeds 1984.

(11) HART, P.E., NILSSON, N.J., RAPHAEL, B. A formal basis for the heuristic determination of minimum cost paths. IEEE Transactions of Systems Science and Cybernetics, Volume SSC-4, number 2, July 1968, pages 100-107.

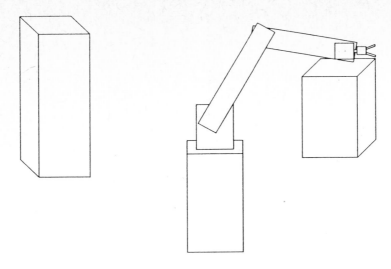

Fig 1   Example of real workspace

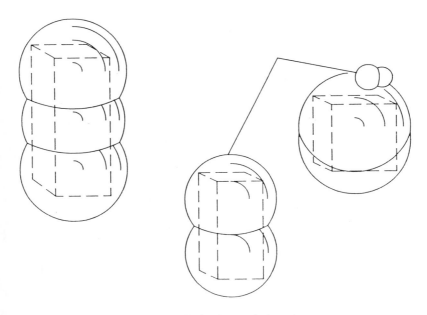

Fig 2   Model of robot and obstacles

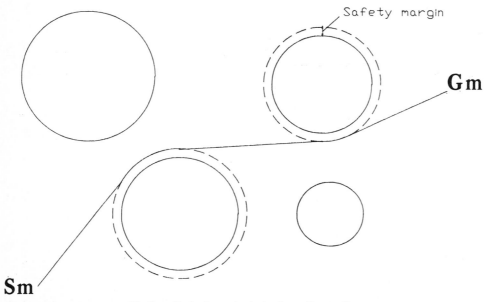

Fig 3   Path through circles from Sm to Gm

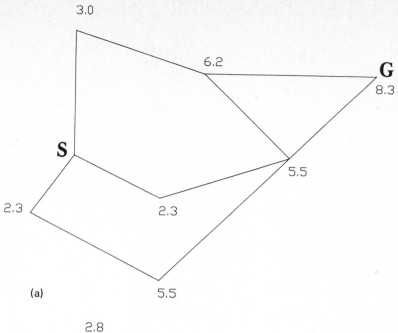

(a)

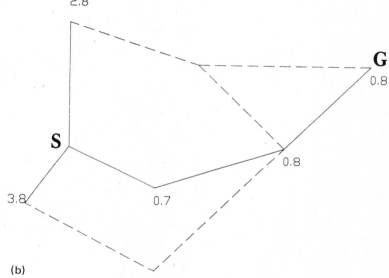

(b)

Fig 4   Path finding strategies
        (a) First search strategy
        (b) Second search strategy

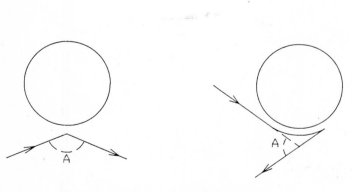

Fig 5   Paths which can be neglected

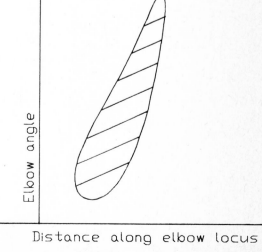

Distance along elbow locus

Fig 6   Obstacle representation in transformed reference
        frame for forearm path planning

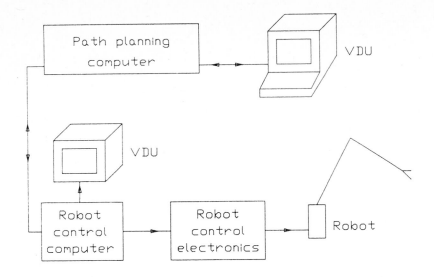

Fig 7   Diagram of system hardware

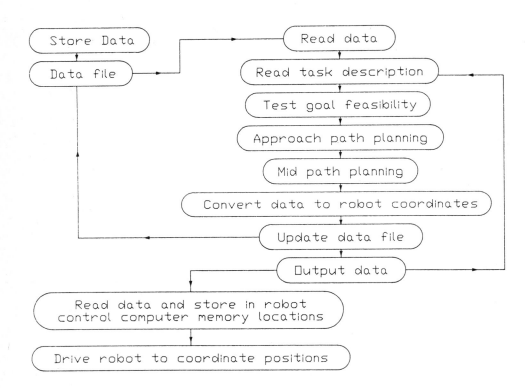

Fig 8   Flowchart of system software

C375/86

Automatically increasing the speed of industrial robots and multi-axis machines

G W VERNON, BEng, J REES JONES, MSc, CEng, MIMechE, MIProdE and G T ROONEY, BEng, PhD, CEng, MIMechE
Mechanisms and Machines Group, Liverpool Polytechnic

Summary

A computer-controlled robot contains all the elements necessary for an autonomous self-experimentation system. The use of this ability is described in minimising task duration while maintaining smooth motion.

The strategy is simple in both design and application, involving incremental reductions in the duration of a given motion. The changes in duration are subject to several constraints including maintenance of axis coordination, specified tracking accuracy and avoidance of actuator saturation or other working limits. Various possible parameters for detecting saturation are considered, and a suitable one is suggested.

The strategy has been implemented on a commercially available robot using a standard microcomputer controller. Conditions for its use are that the robot's response must be repeatable in the short term and that the robot programme is of relatively short duration. The feedback control used in the scheme is of a simple type and there are no extra transducers required.

Introduction

Frequently a substantial proportion of the task time of industrial manipulating robots is made up of non-productive move operations in which the manipulator moves a product, tooling or itself into position for the next process operation. This time adds to the cost of production without adding to the intrinsic value of the product. The objective of the work reported in this paper was to reduce the non-productive time taken up by these move operations without incurring dynamic disturbances or loss of co-ordination.

A point often missed is that a robot control computer is capable of many functions, not simply of feedback control. One of these functions in a (non taught path) robot is the generation of multi-axis motion, in which positions are coordinated with time, which implies the inclusion of velocities, accelerations and higher derivatives that are usually invisible to the user.

Current industrial practice is to set operating speed at some fraction of a conservative maximum speed. This method relies on the operators ability to assess acceptable performance levels by eye and how much the robot can tolerate over sustained periods.

Another approach is to model the dynamics of the robot including its actuators and control system. This can be used to assess the feasibility of a given motion, Rees Jones, Fischer 1984 (1). Alternatively, the model can be used for computing near minimum time paths, (with their time histories). There are several techniques available, such as

Roth & Kahn 1971 (2); Vukobratovic, Kircanski 1982 (3) and Dubowsky, Norris and Shiller (4) 1986. The last of these does consider some limitations imposed by the actuators, but they use only rigid models, and are in simulation, only. The resulting motion designs would require rapid and large changes in actuator torque resulting in transient vibration and high shock loads in a real system. They must also work close to actuator saturation, which can be difficult to quantify within a model. Another disadvantage of these theoretical techniques is that they require iterative and therefore extremely lengthy computations of the dynamic model.

The approach presented here utilises the fact that any programmable industrial robot possesses all the necessary elements of a self contained experimental system. It includes a strategy in which results from a previous run are processed and used to update the command trajectory.

The motion originates from either
(i) A sequence of widely spaced coordinate sets which are boundary conditions of pieces or 'segments' of motion. Added to them may be some of the following constraints: a linear or circular path between succesive sets, in Cartesian space; constant velocity along such paths; maintenance of axis coordination.
or (ii) A complete taught motion, including every coordinate set eg. as from a spray robot teach arm.
This latter case is not of interest

in automatic motion design, as there is no further automation required, in the motion generation process.

If a spatial path is defined, for example a straight line; a velocity or acceleration profile must be defined, along the line, or at the actuator.

An idealised motion design strategy would use a knowledge of the robot's dynamics, the complete workspace, the task and associated limitations in order to optimise for such as task time, or energy usage, component life, minimal vibration etc. Currently this is technically but not economically feasible. The systems actually used incorporate simple motion design rules of thumb. Motion design schemes used by industrial robots are likely to continue to be semi-empirical because of the complexity involved in accurately modelling a robot.

Previous work has indicated that good dynamic results are obtained with fifth degree polynomials, in stop-go-stop motions; Rees Jones, Rooney & Fischer 1984 (5). Lower degree polynomials will generally give step changes in acceleration i.e. instant changes in demand force, exciting vibrations. Significant improvements are not obtained with higher degree laws. Using a simulated second order undamped model response as a performance index, plots of peak error in response to symmetric 3rd, 5th and 7th degree normalised polynomials with zero velocity and acceleration boundary conditions are presented in figure 1. These indicate that the benefits of continuous high order laws, are lost in the discretisation process. Very high sample rates are needed to regain them.

It should be noted that attention is restricted to stop-go-stop motions in this work. This limits potential applications but provides a suitable starting point for further work in this field.

Notation

C - coriolis and centripetal matrix
d1 - initial segment displacement
d2 - final segment displacement
delta - sample interval
F - force vector
Fm - motion frequency; 1 / T
Fn - system natural frequency
Fs - sample frequency
G - gravitational vector
i - sample number; t = i.delta
I - inertia matrix
j - axis or joint number
k - repeat cycle or run number
L - learning gain matrix
q - general joint coordinate vector
r - trajectory 'vector'
r̄ - nominal trajectory vector
Rm - period ratio; Fn / Fm
Rs - sampling ratio; Fs / Fm
t - time

T - motion duration
Wn - system circular natural frequency
ζ - system damping ratio
y - response vector

Increasing Speed

It is assumed that for the purposes of the scheme, the trajectory data is available to us in the form of an integer array of sets of joint coordinates which are equally spaced in time

$$r_{i,j,k}$$

The segment runs from sample number i = istart to ifinish. The corresponding time data is implicit in the sample number i, as t = i.delta. The repeated compression process used can be represented by

$$T_{k+1} = (\text{reduction factor}).T_k$$

One way of achieving this might be to change the sample interval. A problem which arises in doing this is that the amount of change which would be required in the sample interval is widely variable. There is not usually much spare capacity in the central processing unit (cpu) capability to permit a large reduction in sample time. Another approach particularly applicable in the case of a polynomial is to re-compute the function throughout the segment being compressed. If the motion is specified in cartesian coordinates then a series of kinematic inversions is also required. If alternatively, a simple linear interpolator is used, it needs far fewer arithmetic operations and furthermore no kinematic inversion is necessary. Repeated interpolation or extrapolation of the same trajectory will produce accumulative errors. With some care in rounding, these errors are found to be negligible for practical numbers of runs of a given motion.

What is Saturation ?

The output of servo systems change as some (usually dynamic) function of the input. The way in which it changes may be unimportant but there is always a point at which a further change in input has no further effect on the response. In an industrial robot, saturation can occur in a wide variety of elements, for example in an amplifier or a servo valve, limiting current or flow. It will nevertheless manifest itself in the joint displacement response and its derivatives.

Why is Saturation Undesirable ?

Once an actuator is operating in a saturated region the servo sytem can no longer maintain proper control over the motion. this can lead to the loss of tracking accuracy, joint axis coordination and possible collision. As the actuator enters and leaves saturation, there is a discontinuity at some derivative of the force or torque provided by the actuator. The lower the order of discontinuity, the greater the significance of the resultant vibration. This vibration will not only make the task more difficult to perform, but may reduce the components working lives. Working repeatedly in a saturated region may also cause overheating and resultant rapid wear.

One strategy used is to defer output of the coordinate stream when the error exceeds some set limit. This introduces a discontinuity in the displacement curve slope, hence velocity, which in turn will initiate un-necessary vibration.

Detecting the Onset of Saturation

Commonly, robot actuator systems are comprised of either:-

    pre-amplifier ,
    power amplifier &
    permanent magnet d.c. motor
or
    current amplifier ,
    servo valve &
    hydraulic cylinder or motor

Any of these elements can of course reach its working limits. At an amplifier, saturation may be detected from input signal levels. The amplifier's dynamics may be neglected because the rest of the sytems dynamics are much slower, and so the amplifier may be approximated to a constant gain. Saturation in the other items is not so easily quantified. For both electric and electro-hydraulic systems, the steady state speed / torque or force limitations are broadly speaking similar, shown in figure 2. Note that this figure is an over generalisation, but it serves to show the complexity of the saturation bounds. Further problems are that the demand load changes dramatically because of the robot's non-linear dynamic characteristics

$$F(q,\dot{q}) = I(q).\ddot{q} + C(q,\dot{q}) + G(q)$$

Many of these matrix elements have a trigonometric dependency, some of the coordinate velocities are multiplied together, and there is strong joint cross-coupling. These all cause the demand torque function to be non-uniform. In the hydraulic case there is a severe flow coupling by virtue of the common power supply providing flow for all axes simultaneously. An individual axis may draw high flow at low pressure, for example, dropping the system line pressure causing the remaining actuators to be starved of flow. These and other reasons cause the conventional motion

generation strategies to be conservative, sometimes highly so.

Some parameter, is required as an indicator of saturation, as the motion is cyclically reduced in duration. An idealised parameter would have the following properties
 (i)   It must cope with arbitrary motion duration and displacement boundary conditions. (Velocity & acceleration conditions should also be considered in general, but are neglected here.)
 (ii)   It should cope with varying and probably unknown loads at the gripper.
 (iii) The computation required to evaluate it must not be excessive.
 (iv)   It should be repeatable and immune from noise sources such as signal noise, joint   repeatability and   discretisation.
 (v)   It should not be robot specific.
 (vi)   It should be a clear indicator of saturation, i.e. sensitivity should be high, giving a marked 'switching' level.
 (vii) It should be able to detect saturation arising in a small region of the response, or in a single lone axis.

Some of these requirements are conflicting and so a compromise must be reached. It is hypothesised that a single scalar quantity can be found, one per axis, which is normalised to be independent of displacements and time, and satisfies most of the above requirements. Retaining a constant motion law function is later found to simplify the otherwise general nature of the problem.

The progressive onset of saturation is difficult to detect by eye in any one displacement response. For reduction factors close to unity, the change due to saturation between any pair of responses is small. Qualitatively, the robot itself behaves less smoothly and may begin to judder visibly, or overshoot its endpoint. Some typical saturated and un-saturated responses are plotted in figure 3. System units are the 12 bit integer quantities (0-4095) used by the robot to represent the 3 axis displacements. The displacement response as plotted may contain flats (near constant velocity sections) or regions where there has been little change since the last response, albeit the motion duration has been changed. Oscillations may be evident in the overshoot response but are probably small in amplitude and so difficult to quantify for use in any saturation parameter.

Typically the information available on the motion includes the displacement command, its associated response and sometimes, the response velocity. Derived quantities then are the displacement error curves. There is a small change in the proportions of the error curve (figure 3 (b)   ), but it is probably robot and motion specific, and difficult to use without significant processing.
 Of course, as the robot progressively speeds up, the displacement

errors increase in magnitude. The peak and average errors are plotted in figure 4; against the cycle number, hence decreasing motion duration. The starting duration for this particular motion is a highly conservative 3 seconds, with a reduction factor per cycle of 0.95, giving a final deeply saturated duration of approximately 0.5 seconds. The reduction factor value 0.95 results in a nominal 5% tolerance on optimum motion duration. The absolute values of peak or average error are of little use because of their dependence on the specific motion displacements and its duration. The noticeable change in the curve slope however, may be of use.

Initial estimates of motion duration must be conservative to ensure that on the first run, the response is not saturated. Referring to figure 3 (b), error magnitudes have been obtained by differencing already low resolution 12 or 16 bit displacement values. To quantify the slope of the change in peak error, requires the further differencing of successive peak values of error. These peak values are prone therefore to numerical difficulties and resulting inaccuracies in the slope estimation. If instead, an average error is plotted (figure 4), the difficulties although still present, are reduced. Note that because both the peak and average displacement error curves are very similar in shape, the ratio peak / average is of no value.

If a simple model could be fitted to the unsaturated region of the average error curve, departure from it may serve as a saturation indicator. Symmetric polynomials driving second order differential equation models generate average error functions in two components. The gross component can be shown to be given by

$$\frac{2880 . \ddot{?} . (d2 - d1)}{(Wn . T)^5}$$

for a 5th degree polynomial but the transient error component is not of such simple form. If it were negligible, these functions might be of use. Curve fits to the average error were found to be good for functions such as quadratics and cubics in T, also 1/T.

Problems again arise in the attempt to normalise the deviation from the curve. The simple 1/T function suggested a close tie with the flow constraints of this particular manipulator, so was neglected.

Differencing the successive average errors, figure 5 (a), there is a sharp change in this curve. To use it still requires some form of normalisation. It may be possible to use a fundamental time scaling property of manipulator dynamics identified in Hollerbach 1984 (6) but it requires a large amount of computation. If a reduction factor close to unity (say

0.95) is selected, then the ratio of previous average error value over its current value produce a pseudo normalised quantity. Such a quantity is plotted in figure 5 (b).

So far, all parameters tested are based on the displacement. Velocity data is often available, although in this case was estimated using a central differences formula combined with a digital filter to reduce discretisation noise, which occurs predominantly at low speed. Unsaturated and saturated motion velocity curves are shown in figure 3 (c). At lower speed, as expected, the response velocity closely resembles that of the demand.

Saturation causes a deterioration in the velocity curve shape. This can easily be quantified as a shape factor, of peak over average velocity. Using a fixed degree nominal command polynomial with zero velocity and acceleration boundary conditions the ratio of peak to average velocity can be shown to be constant. It is independent of both the motion duration and the displacements. For a fifth degree law it is 15/8. If the ratio for the actual response velocity is divided by this factor, the resulting quantity should be close to unity at low speeds. As the velocity curve flattens, the peak / average ratio reduces detectably (figure 6). A level can then be selected, corresponding to saturation, and used as a 'cut off' limit to stop any further reduction in motion duration.

Advantages with this ratio are that the peak velocity region is also the most likely region to velocity saturate first, so its sensitivity to localised saturation should be good. It will also be less susceptible to displacement repeatability variations from cycle to cycle, because it is only based on differences in displacement values, from the current cycle. This means it can be evaluated on the first run as compared with two or three runs needed with other quantities suggested. Without a good feedback controller, it may though suffer from more serious coupling and gravitational components than was experienced in these tests, leading to velocity curve distortion even in the un-saturated response. More work would be required to assess the effects.

Note that saturation is detected only after it has occurred, and so the motion must be re-cast, being expanded to a previous acceptable level. This is carried out for the additional purpose of permitting motion tuning for increased accuracy. There needs to be some 'reserve' in actuator performance, to enable it.

Maintenance of Tracking Accuracy

As stated earlier, loss of tracking accuracy can have serious consequences. Further, the motion following this segment may have need for high tracking accuracy at its commencement.

There are many feedback control schemes designed to cope with the problems of robot control, so it will not be addressed here. There is an additional method of attack, to improve tracking by modifying the motion before it is fed to the robot's servo system & actuators.

So called dynamic tuning of motion laws has been with us for many years, mostly from the field of cams.

Normally the process is to feed the nominal plant command into a dynamic model in place of the response, and a new, dynamically tuned motion command is computed based on it. Although this is carried out in some schemes, the dynamic model complexity in the robot case makes it an expensive solution. The scheme of Goor 1985 (7) does produce a practical system for robot motion tuning. This requires the evaluation of several derivatives of the motion, but not a full dynamic model.

The scheme outlined here is presented in more detail in Vernon, Rees Jones & Rooney 1986 (8), but is straightforward in its implementation. It is derived from a linearised state space model. This only needs to be valid for the duration of a single sample, and none of the dynamic coefficients need to be evaluated.

The accuracy increasing algorithm is recursive in its form. A conventional digital feedback controller manipulates the system input, based on previous run information. If this particular piece of motion has been run in exactly the same way, the cycle before, then the previous response can be used to augment the current discrete motion data in the following way

$$r_{i,j,k+1} = r_{i,j,k} + L.(\overline{r}_{i+1,j} - y_{i+1,j,k})$$

Although the L matrix for a perfect solution is highly complex, in practice it can be adequately approximated to a near identity matrix.

The algorithm cannot be used to process the entire motion in a single shot. By using a 'window' over the motion segment, comprising an exponentially decaying factor in front of the L matrix, tracking errors have been minimised in as few as 8 cycles. Close to the sample being tuned, the factors value is unity, and it decays away with increasing time.

The dynamic tracking attained is very close to the static accuracy in our particular system. Noise present in our system led to the use of a digital filter applied to the component added to the motion.

Figure 7 shows the combined effects of the use of both the speed increasing and accuracy increasing routines, on a typical motion.

Implementation

The robot on which the scheme was tested was a 4 axis R-R-P-R configuration (R - revolute, P - prismatic joint) of which only the three major spatial axes were used. The complete motion generation & control scheme has been set up on an i8086 based commercial micro-computer. It is outlined in figure 8. The proprietary analog interface contains three 12 bit digital to analog converters, one 12 bit analog to digital converter with 32 channel multiplexer and a programmable real time clock. The sample interval was set at 10 ms; a maximum value of the robot's fundamental frequency being around 16 Hz. The three major axes were driven using simple error proportional control, with a stabilising pressure feedback on the swing and vertical axes, all axes being hydraulic. The control was analog in form to reduce the cpu loading, permitting the development and testing of various strategies as outlined. Part of the reason for the cpu load was a 6 ms overhead per sample due to the IEEE488 communications and various analog interface delays. In a final hardware implementation, currently nearing completion using an M68010/VME based system and customised interface, this overhead has been reduced to around 80 micro seconds.

Control gains were empirically maximised to yield the highest tracking accuracy consistent with stability requirements. The servo valves were nominally flow proportional and supplied from a common fixed displacement hydraulic power supply. Displacement feedback was provided by 0.1% linear potentiometers.

The various strategies were implemented as separate procedures in Pascal, which could be switch selected by changing the contents of a data file. The basic motion used comprised five segments, of which only one, typical segment, is presented.

The system has been designed to generate motions of arbitrary numbers of segments and axes. To this end three programs have been developed, one for arbitrary polynomial motion generation, a second for computing the motion data and finally a 'run' program which controls and supervises the robot in real time. The motion development process can be tedious for the user and so an automated version has been produced in the form of a crude computer language compiler. More control can be exercised over the motions used than in a typical robot language.

Concluding Remarks

It is a simple process to cyclically compress a motion, and it has been demonstrated that the achievement of near optimum stop-go-stop motions is easily realisable.

Anticipating the ideal of autonomous plant, other more complicated motion segments need to be considered, for example motions between fixed and moving constraints. Here the technique presented for characterisation of saturation will, in general fail.

The most suitable parameter found in this study is the ratio of peak to average velocity. Although not ideal, it satisfies most of the requirements for such a quantity. At low speed it remains relatively constant, resembling the 5th degree symmetric polynomial command velocity and so has a value around 15/8. As distortion arises in the velocity response, due to saturation, its value drops. The progressive increase in tracking error which accompanies the process is tackled using a self learning algorithm in which previous response data is used to augment the motion command.

The complete technique is applicable to other multi-axis machine tools under the appropriate conditions.

A more suitable parameter is yet needed to cater for the diversity of general motions.

Acknowledgments

This work was supported by research grant GR/D/24081 under the ACME Directorate of the Science and Engineering Research Council. Thanks are also due to Mr S.A.Caulder who constructed and maintained various items of equipment on the project.

References

1. Rees Jones, J., Fischer et al, "Final Report, Dynamically Optimum Performance of Manipulators", SERC Grant GR/B/51840, July 1984.
2. Kahn,M.E., Roth,B.,"The Near Minimum-Time Control of Open-Loop Articulated Kinematic Chains", ASME Trans., J of Dyn. Sys., Meas., and Control, September 1971, pp164-172.
3. Vukobratovic,M., Kircanski,M.,"A Method for Optimal Synthesis of Manipulation Robot Trajectories", ASME Trans., J of Dyn. Sys., Meas., and Control, June 1982, vol 104, pp188-193.
4. Dubowsky,S., Norris,M.A., Shiller, Z. "Time Optimal Robotic Manipulator Task Planning", 6th CISM - IFToMM Symposium, Ro.man.sy - 86, Cracow, Poland, September 1986.
5. Rees Jones,J., Rooney,G.T., and Fischer,P.J.,"Dynamically Smoother Motion for Industrial Robots: Design and Implementation", UK Robotics Research Conference, London, I.Mech.E., December 1984.
6. Hollerbach,J.M.,"Dynamic Scaling of Manipulator Trajectories",ASME Trans., J of Dyn. Sys., Meas., and Control, March 1984, vol 106, pp102-106.
7. Goor,R.M.,"A New Approach to Minimum Time Robot Control",General Motors Research Laboratories,Warren, Michigan, document GMR-4869, November 1984.
8. Vernon,G.W.,Rees Jones,J., and Rooney,G.T.,"Dynamic Command Motion Tuning for Robots:- A Self Learning Algorithm",6th CISM - IFToMM Symposium, Ro.man.sy - 86, Cracow, Poland, September 1986.

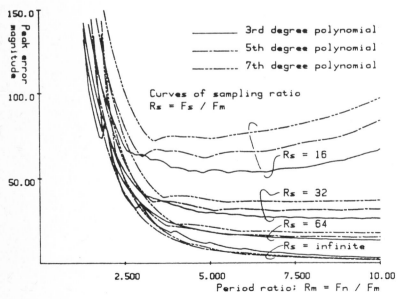

Fig 1   Peak displacement error as a function of period and sampling ratio

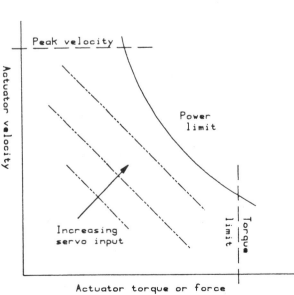

Fig 2   Typical actuator steady state characteristics

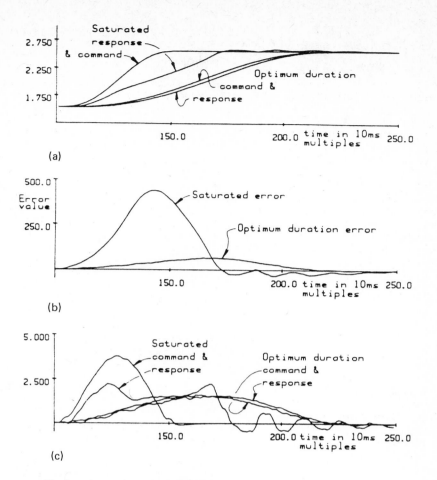

Fig 3   Typical saturated and unsaturated motions
          (a) Displacement command and response for axis 2 (typical)
          (b) Associated displacement errors (typical)
          (c) Numerically established velocity curves for axis 2 (typical)

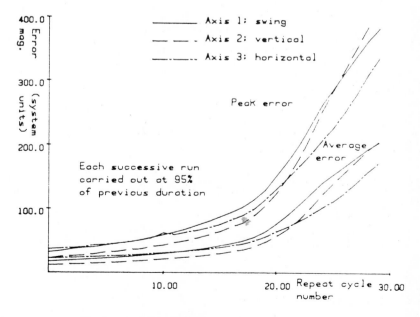

Fig 4   Increase in displacement errors during motion
         compression

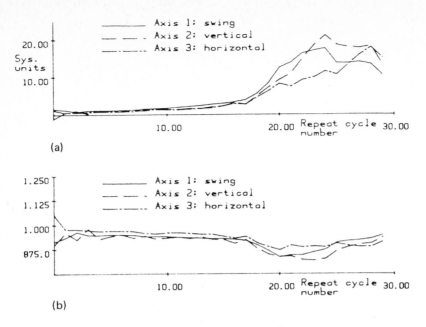

(a)

(b)

Fig 5   Displacement error-based saturation parameters
(a) Change in average error after each run
(b) Ratio of last average error/current average error

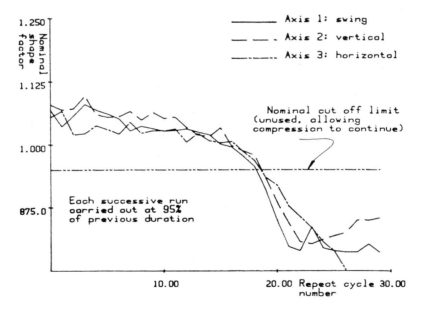

Fig 6   A velocity shape factor as a saturation parameter

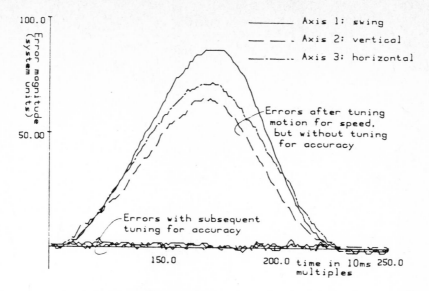

Fig 7   Displacement error curves in tuning for accuracy

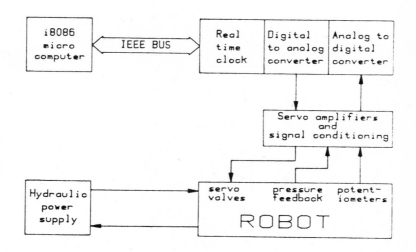

Fig 8   A schematic representation of the system hardware

A knowledge-based approach to robotic assembly

K SELKE, BSc, **K G SWIFT**, BSc, MSc, CEng, MIMechE, MIProdE, **G E TAYLOR**, BSc, MSc, PhD,
A PUGH, BSc, PhD, CEng, FIEE, FIERE, FRSA, **S N DAVEY**, BSc and **G E DEACON**, BSc
Department of Mechanical Engineering, University of Hull

Abstract

In order to increase the efficiency and applicability of robotic assembly systems it is important to monitor assembly progress, identify errors and enable the robot to recover from an error situation. The introduction of Artificial Intelligence techniques in this area is aimed at improving the aid given to the robot by the control computer.

A knowledge-base has been created which encapsulates expertise on error recovery. The knowledge can be used to specify, for particular sensor data associated with assembly operations, those actions which would facilitate recovery without operator assistance.

This system is applied to robotic assembly problems provided by the industrial partner Electrolux plc in Darlington and its performance is discussed in this paper.

1 INTRODUCTION

An important factor in limiting productivity in manufacturing processes is the restricted use of automation in assembly. Whereas dedicated machinery has been proven to be successful with large volume products, utilisation of robots and their associated expensive equipment decreases drastically with the batch size. The use of sensors and their ability to detect changes in the environment brought about by changing products can contribute significantly to the reduction of this problem and thus spread the cost of equipment over a sufficiently large number of batches to make automation viable.

Such a strategy appears to be most attractive, but problems encountered in practical implementations have highlighted the need for a systematic approach to the coordination of sensing and manipulation during assembly. Experiences gained at the University of Hull whilst studying generic assembly problems [1] are largely matched by those gathered whilst using robots on the production line at Electrolux [2]: Substantial efforts have to be applied to each batch in order to achieve satisfactory performance levels, making robotic assembly unsuitable for smaller batches.

This paper presents the approach chosen by researchers at the University of Hull, combining academic and industrial experiences to result in flexible automated assembly. It is suggested that the creation of a sensor-rich environment providing complementary information from different sensors enhances the knowledge of a cell controller about the assembly under progress. Flexibility is increased by matching sensory systems to essential assembly processes as well as interpreting the sensor signals intelligently. Thus the cell tries to determine the assembly success by relating sensor signals to those experienced previously in a similar context. Consequently, Artificial Intelligence techniques are well suited to deduce error recovery strategies from the data available in the cell.

An example is given by implementing a specific product assembly from the Electrolux range (FLYMO lawn mower) in a demonstration cell at the University of Hull.

2 SENSOR-GUIDED GENERIC ASSEMBLY

In order to choose sensors for the cell, the assembly process itself has to be investigated. Three distinct stages can be identified (see figure 1) during the addition of a new component to an existing structure [3]:

(a) Feeding
   A new component has to be brought within the reach of the robot, and a large amount of knowledge already exists for this purpose from conventional automation. However, such techniques are only economically justifiable if the feeders themselves can be made multipurpose and important results have been published in this area [4,5,6,7].

(b) Component Transport
   Once the new component has been fed into the cell, it has to be moved from the feeding outlet to the point of assembly. Although the manipulation of the new piece part is not without influence upon the success of assembly, the emphasis in this stage is directed towards aspects of manipulator control and gripping. A secure grasp, collision avoidance and minimal transportation times are the dominant issues here.

(c) Component Mating
   The third and final stage is the assembly process itself, i.e. the joining of the components. It is characterised by the forces and torques resulting from the

components touching and has been the subject of intensive research [8]. Results obtained from the analysis of part mating have led to successful applications of active and passive compliant devices.

Whereas such devices are well able to compensate for misalignments, they either do not provide additional information to the cell controller or the information is seen in isolation, leaving the eventual outcome to assumptions, rather than explicit determination. The information content in an assembly situation has three distinct phases (see figure 2) and their associated parameters can be monitored with sensors, so that the cell controller is able to make decisions derived from the current process itself. The three important phases for gathering sensor information are considered in turn:

(a) PRE Mating
Before the mating of the parts can be attempted, the relative location between the new part and the subassembly has to be minimised. Usually, it cannot be measured directly but has to be calculated from absolute location of the component and the absolute location of the subassembly. Because errors remain in measurement systems and positioning devices, nonlinearities and other shortcomings are accentuated, and since clearances are generally smaller than the sum of errors encountered, the sensor information cannot solely be relied upon at this stage.

(b) DURING Mating
The misalignment resulting from the remaining errors produces initial contact forces, and by minimising these, correct alignment is achieved. Thereafter the parts are sliding with respect to each other, so that a force has to be exerted along the insertion direction to overcome resistance. Depending on complexity, the sliding motion may be composed of several segments. The mating is terminated either by reaching a default insertion depth or by experiencing a change in force, as may be the case for a snap fit, or by encountering both as in the case of an endstop.

(c) POST Mating
After the manipulation of the component has finished, it is released and becomes an integral part of the subassembly. An explicit decision needs to be taken in this third phase, evaluating the sensory information with respect to assembly success. Sometimes, the sequential nature of the process may provide that information merely by entering POST mating, but this phase also has scope for the incorporation of final functional testing and inspection facilities.

3 SENSING GENERIC PARAMETERS

Mating features, i.e. the precise regions of the component which are important for assembly, are of course highly product dependent. This poses problems in the choice of sensing systems which can be used economically for the large variety of product components. However, initial investigations [9,10,11] indicate that the majority of components have assembly features which can be grouped into different classes, and features contained in each of those classes can then be identified with the relevant sensors. Work on the identification of these features is continuing.

3.1 The Hardware Structure Of The Cell

Although the classes of assembly features reduce the number of sensors necessary in a cell, flexibility has to be maintained in the ability to change the sensing and actuating devices so that new classes can be accommodated. The need to uphold software and hardware modularity for the implementation of a potentially complex structure resulted in a cell communication and control system called ROBUS [12]. Physically, ROBUS is a parallel bus, on which different devices may exchange information independently of their functionality. A supervisory computer (master) monitors and initiates sensor or manipulator actions and controls the messages passing from one device attached to the cell (slaves) to another, controlling the cell in real time. The ROBUS definition poses no demands on the local intelligence of a sensor or actuator, but as there are usually microprocessors attached for a variety of control and processing functions, commands can be passed in tokenised form, hence reducing the amount of data on the bus.

3.2 The Software Structure Of The Cell

In order to initiate the gathering of sensor data and relay information from one slave to another, the master needs a certain amount of knowledge in order to perform the required sequence of actions for correct assembly. This information is produced during the teaching phase (see figure 3) and has the form of a data file containing the assembly instructions specifically for one product. Teaching is currently carried out manually, but may equally well be done textually [13], thereby allowing the use of CAD facilities to generate the data offline.

Apart from these action sequences, this product file contains the sensory requirements for each of the generic assembly processes and the master controller can monitor the relevent tolerances. Thus the presence of an error can be detected by the real time controller, if the measured values do not match those expected.

4 A KNOWLEDGE-BASED APPROACH

By passing the measured sensor data back to the problem solver (see figure 3), a continuous updating can be achieved, and by applying an appropriate statistical filter (eg Kalman filter) [14], the current values are seen in the correct context of sensory feedback experiences. To some extent the error detection is duplicated, but whereas the master finds the predicted errors, the problem solver works in the background and detects trends and context dependent errors.

Hence 'interrupts' work in two ways; an immediate error condition initates an action in the problem solver, whereas others (like gradual changes) can alter the current actions of the cell controller. Detected error conditions can be related to the sensor signals as well as the particular place in the sequence in which they occurred. Primed with this information, the problem solver is able to access the knowledge-base containing the rules governing the generic assembly processes. It can now deduce from these rules and the current state of the cell an appropriate action.

Currently, the cell does not contain automatic error recovery generation, although such schemes have been proposed [15]. Usually, error recovery is very product dependent and teaching the recovery actions utilises the knowledge of the operator. Some conditions can be foreseen and hence included in initial recovery strategies. Consequently, errors have a seriousness factor assigned to them, where the degree of seriousness invokes appropriate actions. For example, a slight error may initiate a servo-loop for compensation, whereas moderate errors bring the cell to an orderly halt and ask for operator assistance. Very serious errors result from hardware failures and are currently implemented in the demonstration cell as time-out conditions.

Although at first glance, teaching of error recovery appears to restrict the flexibility, the shopfloor experience at Electrolux has shown that most of the errors are of a certain type and appear regularly, requiring the same intervention by the operator. Thus, together with the predictable error conditions, most reasons for assembly failure will be eliminated very rapidly.

Furthermore, most products are assembled correctly (> 80%) at the first attempt (see figure 4), and lengthy checking and sensory activity increase the cycle time unnecessarily. In order to keep the time for sensory actions to an acceptable level, it is important to introduce them only at key stages. These, in turn, are a result of the confidence which the cell has built up over a number of product assemblies and the value added with each component. Wherever possible, sensor actions can be distributed, with reference to sensory phases in assembly processes, so that they either are executed 'on the fly' or in parallel with activities elsewhere in the cell.

5   IMPLEMENTATION

The product chosen for the demonstration cell has been selected from the range of Electrolux Products. The FLYMO lawn mower is a typical example for the assembly of engineering products. The dominant direction of assembly is vertically downwards, and the number of components involved in the selected subassembly is twelve (see figure 5). Their feeding has been simulated by providing only the 'front end' of the feeders, so that small runs (about a dozen) can be implemented for evaluation purposes.

5.1 Generic Example

One of the components which presented considerable difficulties for implementation on the shopfloor was the bush (see figure 5). Hence it is given here as a specific example, showing its relation to a generic assembly process. Note, that neither the orientation of the bush itself nor that of the motor shaft is easily determined in advance. However, the misalignment results in specific forces and torques encountered at the initial contact, and because an additional degree of freedom is provided by the motor itself, it can be exploited during the alignment procedure for faster response. Figure 6 shows sample generic rules from the system's knowledge-base.

5.2 Artificial Intelligence Techniques

The interest in Artificial Intelligence techniques arose from the idea that it should be far easier to tune a rule-based system to the requirements of a cell controller than a conventional 'fixed' program. A knowledge-based system is well suited to the application of the specialist knowhow built up in industry and at the University over a number of years. The clean separation of the expert knowledge from the inference mechanism that uses it enables such systems to be readily updated, with knowledge stored in a very transparent form.

In this way, the complex and varied data coming from the sensory systems can be interpreted and related to systematic failure recovery strategies embedded within the rules of the knowledge-base. In the case of a new error situation being identified, the appropriate recovery routine can be added to the body of knowledge, for use as required at a later stage.

Most of our work is carried out using the logic programing language PROLOG [16]. PROLOG provides a standard inference mechanism for rules (clauses) with multiple solutions, but there are no inbuilt mechanisms for presenting inferences or giving explanations. However, since PROLOG is a powerful general purpose language with meta-rule primitives, it can be used to program the extended mechanisms required.

6   CONCLUSIONS

By considering the process of assembly in its constituent parts, a suitable structure for the implementation of a flexible assembly cell has been proposed. It consists of an adaptable hardware support, in which sensors and actuators are easily integrated. The software is divided into two main parts, one for the real-time control of sensor and actuator interactions and the other for interpreting the sensor information in the correct context. The latter part is highly suitable for software tools from the area of Artificial Intelligence. Finally, because the knowledge about generic assembly operations is available to the cell through a continuously updated knowledge-base, it is possible to consider error recovery strategies, and although the generation of recovery routines is not automatic, the industrial input has given valuable insight into the importance of the kinds of errors occurring. Manual teaching for corrective actions is thus seen to be a feasible alternative. Work is continuing on the identification and classification of generic assembly types, and is aimed at providing a systematic approach to the choice and usage of sensing methods in flexible assembly stations on the shop floor. Consequently, the effort of setting up a cell will be reduced and thus the cell itself becomes available for smaller batches, providing a utilisation which is at least comparable to other assembly equipment.

7 REFERENCES

1. Hill,J.J., Burgess,D.C., Pugh,A., "The vision-guided assembly of high-power semiconductor diodes", Proc. of 14th ISIR, Gothenburg, Sweden, October 2-4, 1984, pp 449-459

2. Leete,M.W., "Integration of robot operations at Flymo - A case study", Proc. of 6th BRA Annual Conference, May 16-19, 1983, Birmingham, UK, pp 113-124

3. Selke,K., Pugh,A., "Sensor-guided generic assembly", Proc. of 6th RoViSeC, June 2-4, 1986, Paris, France

4. Cronshaw,A.J., Heginbotham,W.B., Pugh,A., "A practical vision system for use with bowlfeeders", Proc. of 1st International Conference on Assembly Automation, March 1980, pp 265-274

5. Swift,K.G., Dewhirst,R.J., "A laser electro-optic device for the orientation of mass produced components", Optics and Lasers in Engineering, 4, 1983, pp 203-215

6. Browne,A., "A trainable component feeder", Proc. of 5th International Conference on Assembly Automation, May 1984, pp 85-93

7. Redford,A.H., Lo,E.K., Kileen,P.J., "Parts feeder for a multi-arm assembly robot", Proc. of 15th CIRP International Seminar on Manufacturing Systems Assembly Automation, June 1983, Massachusetts, pp 118-125

8. Nevins,J.L., Whitney,D.E., "Assembly research", Automatica, Volume 16, No 6, 1980, pp 595-613

9. Swift,K.G., "A system for classification of automated assembly", M.Sc. Thesis, University of Salford, 1980

10. Williams,A.M., Walters, P.E., Ashton,M., Reay,D., "A flexible assembly cell", 6th International Conference on Assembly Automation, May 14-16, 1985, Birmingham, UK, pp 57-66

11. Groover,M.P., "CAD/CAM: Computer-aided design and manufacturing", Chapter 12, Publ. Prentice/Hall International, 1984

12. Taylor,P.M., Stubbings,C.A., "Software and hardware aspects of a flexible workstation for assembly tasks using sensory controls", Proc. 2nd IASTED International Symposium on Robotics and Automation, Lugano 1983

13. Popplestone,R.J., Ambler,A.P., Bellos,I., "RAPT: A language for describing assemblies", The Industrial Robot, September 1978, pp 131-137

14. Johnson,D.G., Hill,J.J., "A Kalman filter approach to sensor-based robot control", IEEE Journal of Robotics and Automation, Vol RA-1, NO 3, September 1985, pp 159-162

15. Lee,M.H., Hardy,N.W., Barnes,D.P., "Research into automatic error recovery", Proc. of Conference on UK Robotics Research, Inst. of Mech. Eng., London, December 1984, pp 65-69

16. Roussel,P., "Prolog: Manuel de Reference et d'utilisation", Universite Aix-Marseilles, Luminy 1975

8 ACKNOWLEDGEMENTS

The support of the SERC under Grant Number GR/C/88776 is gratefully acknowledged. Also, the Department of Cognitive Systems at the Central Research Laboratories of Thorn EMI in Hayes under Dr. John Parks has provided valuable insight into the subject.

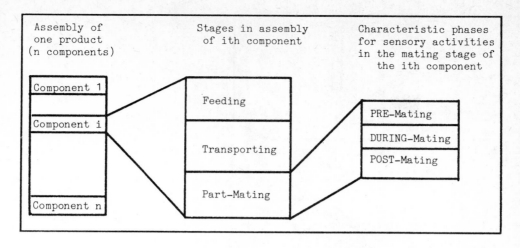

Fig 1   Stages in the assembly of a component

Characteristic phase for sensory activity	Assembly information	Physical parameters
PRE-Mating	Relative Location	Location of assembly point
		Location of new component
DURING-Mating	Initial Contact	Contact forces
		Contact torques
	Sliding Contact	Assembly direction segments
		Assembly forces for segments
	Final Contact	Insertion depth
		Change in force
POST-Mating	Assessment Function	Functional testing
		Quality testing
		Assembly success

Fig 2   Phases for sensory information

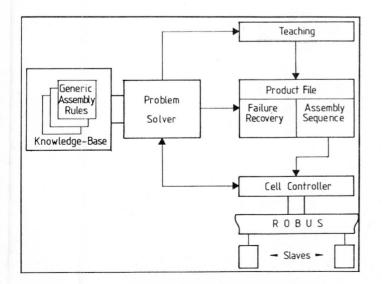

Fig 3   Structure of the knowledge-based assembly cell

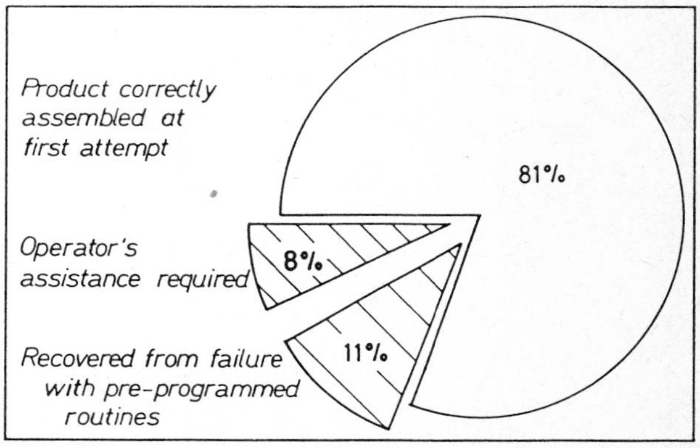

Fig 4   Failure distribution in robotic assembly

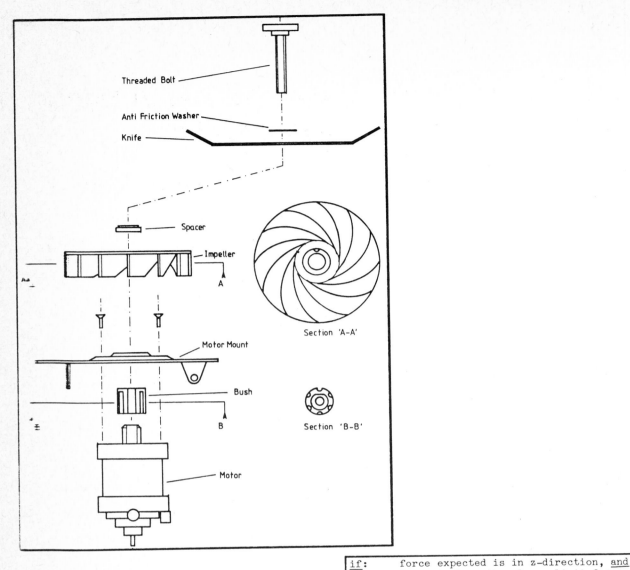

Threaded Bolt

Anti Friction Washer

Knife

Spacer

Impeller

A

Section 'A-A'

Motor Mount

Bush

B

Section 'B-B'

Motor

Fig 5   Flymo lawnmower sub-assembly

```
if:     force expected is in z-direction, and
        force is less than lower limit force, and
        component is in gripper
then:   possibly
        situation: components are correctly aligned,
        action:    begin insertion process

            ..................

if:     force expected is in z-direction, and
        force is between lower and upper limit forces
then:   certainly
        situation: bush touching - but not aligned
        action:    drive motor shaft

            ..................

if:     force expected is in z-direction, and
        force is larger than upper limit force, and
        component is in gripper
then:   possibly
        situation: initial contact error - axial
        misalignment
        action:    investigate (x,y) forces

            ..................

if:     torque expected is in x-y plane, and
        torque is between upper and lower limit torques
then:   certainly
        situation: satisfactory pre-mating
        action:    begin insertion process

            ..................
```

Fig 6   Generic assembly rules for the bush insertion

C367/86

A system for the design and evaluation of highly automated batch manufacturing systems

R BELL, BSc, MSc, PhD, DSc, CEng, MIMechE, MIEE, **E A ROBERTS**, BSc, MA, PhD, CEng, MIMechE,
N SHIRES, BSc, AMIMechE, **S T NEWMAN**, BSc, AMIProdE and **R KHANOLKAR**, BTech, MSc, CEng, MInstE, AMIMechE
Department of Engineering Production, Loughborough University of Technology

SYNOPSIS This paper describes a coherent suite of data driven software for manufacturing system design and assessment mounted on a parallel processing dedicated workstation. Two major phases are recognised. Firstly an evaluation phase provides rapid appraisal of system performance using average measures based on queueing theory. Secondly an emulation phase generates detailed dynamic statistics. A scale layout is input and material flow paths defined. Specific demand patterns are emulated under system constraints and with desired operating rules. Emulation modules are processed in parallel.

1. INTRODUCTION

Modelling, evaluation and the assessment of advanced manufacturing system performance is a complex problem and currently requires a large investment in capital cost and manpower to solve. The tools for a fluent cost effective assessment are clearly not available at this time. This paper is a progress report on tools being developed for the design and assessment of manufacturing systems at Loughborough University of Technology.

2. SYSTEM OBJECTIVES

The major objective of the work is to produce and integrate a suite of design tools for advanced manufacturing systems which can be used by manufacturing design engineers. This is being achieved by having a design system which is data driven and is user friendly using a dedicated computer workstation; by making modelling and simulation transparent to the user; and by providing rapid feedback of system performance parameters from both steady state and dynamic assessment methods.

3. MODELLING AND SIMULATION IN FMS

The high capital cost and complexity of work are now accepted features of the design of flexible manufacturing systems. The design study requires a large commitment of manpower and skill for the correct specification and integration of manufacturing elements into the system. Hence the use of computer design tools to aid the designer is becoming of greater significance [1]. Requirements for the design of advanced manufacturing systems are that modelling methods need to be extremely flexible and that they enable evaluation of alternative systems with different modelling criteria.

Modelling methods which are typically available for use in manufacturing systems can be categorised into the following areas: mathematical and queueing theory models, physical models, graphical simulation and discrete event simulation models. [6] Mathematical and queueing network models are widely used because they have proven to be an effective tool for the initial sizing and balancing of manufacturing systems. However, most of these models make major assumptions, (such as exponential service rates, FIFO queues, infinite part buffers etc) and provide steady state outputs of station utilisation, transport utilisation, queue lengths, production throughput etc. [7].

Physical models of manufacturing systems are usually built using modelling kits, which provide scale models of system elements to be used in the final manufacturing system [3]. These types of models can be used to test system software, before final use on the actual system.

Computer based graphical simulators are used for modelling and evaluation of industrial work cells and manufacturing systems. The main applications are in the design and layout of work cells, the simulation of robot movement patterns and collision detection. [4]

Discrete event simulators, allow systems to be modelled at a greater level of detail compared to mathematical modelling. A computer program is produced which describes a system in terms of the objects contained, which are called entities, and the states through which these objects pass as simulation proceeds. Four main approaches to program writing are available as described in [6]. Prior to the start of this research programme, considerable effort and time was required to build a model, particularly of a complex manufacturing facility. Recently efforts have been made to reduce this time by providing program building front ends although it is not unusual for some program level work still being necessary for the modelling of complex systems. In addition, computer CPU time to execute simulation runs was relatively high, particularly for complex models. Again, in many cases, specific manufacturing schedules could not be easily used within the model, sampling techniques being used to generate this type of information.

Recently some attention has been paid to these problems and there are reported systems available specifically for the simulation of manufacturing systems, e.g. MAST (5). To some extent, these systems, which have been developed during a similar time period, have paralleled the work reported in this paper but use conventional computer facilities.

4. SYSTEM OVERVIEW

The work described in this paper is the result of a collaborative research project involving Loughborough University of Technology, Baker Perkins Ltd of Peterborough and Normalair Garrett Ltd (NGL) of Yeovil. This project has been funded by the Science Engineering Research Council, Baker Perkins Ltd and Normalair Garrett Ltd to produce and use computer based methods to assist the designer of advanced manufacturing systems.

A prototype computer based advanced manufacturing system design aid has been produced as a result of this project.

Areas of work within the design aid include:

(i)   Interactive manufacturing system data input and manipulation using a database management system.

(ii)   Menu driven software for interactive systematic specification of manufacturing systems.

(iii)   Automatic configuration of data for mathematical modelling and detailed simulation from (i) and (ii).

(iv)   Mathematical queueing modelling for the initial evaluation of manufacturing systems.

(v)   Interactive graphical layout of a manufacturing system to determine the floor plan.

(vi)   Software for the entry of layout data on transport routes and pallet movement. This data is necessary for the emulation phase and for providing animated graphical output.

(vii)   Manufacturing system emulation to provide detailed modelling and determination of dynamic values of system parameters.

(viii) Development of a novel approach to simulation by using multi-processing, on a dedicated parallel processing micro-computer, to underpin the design system.

Areas (i) to (iv) comprise of an initial evaluation stage (see Figure 1) which by using an interactive "softkey" approach as successfully used in MDI systems, collects and collates system data to specify a manufacturing system, and uses a mathematical model such as CAN-Q, developed by Solberg, Purdue University [2] to obtain approximate performance characteristics.

Detailed modelling of manufacturing systems, which in this research is termed emulation, embraces all areas except (iv). Emulation requires a plant layout with detailed information on transport and workpiece handling. The emulation is automatically configured from the system specified in (ii), and run using parallel processing methods to obtain dynamic system output.

5. EVALUATION PHASE

The evaluation software is divided into three main modules, and comprises of (i) data input modules, (ii) specification build module and (iii) specification print module.

(i)   Data Input Module: This aids the user in the interactive entry of data using a soft key dialogue. It is currently a prototype system using manual input, but with suitable file interfaces it is intended that data already available in company computer files will be transferred automatically.

(ii)   Specification Build Module: This module allows the designer of manufacturing systems to interactively create and edit a detailed specification of a manufacturing system. Individual elements of a manufacturing system, which have been pre-defined as machine stations, parts and tools, pallets, part buffer stores and material handling systems are specified. Data relating to specific elements is obtained by the system from the data base set up in module (i).

(iii)   Specification Print Module: This module enables listings or prints to be made of the FMS specification defined in the specification build module.

At present the CAN-Q mathematical queueing model [2] is being used to initially assess steady state performance values of defined systems. Information from the specification of the manufacturing system together with appropriate data from the database is automatically accessed by an interactive program to automatically determine the input requirements for the CAN-Q model. Input information requirements are number of pallets within the system, number of machines, number of servers at each machine, probability of a visit to a machine by a pallet, average processing times of the machines, number of transporters, probability of a transporter being used and average transporter time to a station. This input data is then run automatically through the model to produce steady state results.

A comparative assessment of the suitability of fast approximate modelling methods for the initial evaluation of manufacturing systems is currently being undertaken.

6. EMULATION PHASE

The term emulation is being used in the work at Loughborough to describe a model which imitates a system in regard to all the variables which it is possible to measure while using a particular computational vehicle. The major difference between emulation and simulation lies in the level of detail contained within the model. As indicated earlier, simulation may well contain assumptions on demand patterns, processing times etc., which may be sampled from distributions. Emulation uses actual values for such parameters.

This phase of design includes the configuration by data of a set of software modules each of which emulates a particular set of permanent entities within a manufacturing system. For the purposes of this research, these have been defined as previously, part storage buffers, loading and unloading buffers, machines, tooling and automatic guided vehicles (Figure 2).

Each of these entities has appropriate characteristics which are parameters of the entity. The activities which relate to these are also pre-defined and parametric. An example of this is a rotary buffer which may rotate clockwise, anticlockwise or in both directions and at a particular speed. The buffer loading control may include algorithms to place particular pallets in certain positions or, for example, in the closest free position. The logic used to simulate the action of a particular buffer type and speed is pre-defined and is included in the activity of that entity when the appropriate parameters are chosen for that entity. This approach frees the designer from the construction of activities and events, thus providing a system for reducing the designer involvement in simulation code.

The configured modules are linked together by the system software to create a model which emulates a particular manufacturing system. The model is executed for a suitable time period to generate a file of data which can be analysed later to create statistical data and to control an animation of a plant layout of the manufacturing system.

Information from the designer's initial specification is automatically generated to satisfy the input data requirements of the emulator. This is achieved by running a suite of programs which link the specification to the data base to create the machine data, buffer data, station data and automatic guided vehicle data. Information concerning physical movement of entities of the manufacturing system is generated in a number of steps. The first step is to draw a plan of the static elements of the layout using the DOGS computer aided design graphics system. Once complete, an IGES (Initial Graphics Exchange Specification) file is produced of the layout, using an option within the DOGS systems. A computer program is then used to interpret the IGES file and generate the information required to draw the layout on the particular output terminal used, an IMLAC vector refresh terminal. Using this definition of the layout a number of computer programs are run with designer interaction, which will create the data files for the routing of the automatic guided vehicles, the flow of pallets within the system, and the operation of all the buffers within the system. Use is also made of this information in displaying animated output of the system when in operation.

7. COMPUTER HARDWARE

At the start of the project, it was clear that the level of detail aimed for in the emulator phase would require considerable computational power. Furthermore, the nature of the problem indicated that it was possible to subdivide the project into modules which could be processed concurrently in parallel.

After considering a number of alternatives, including LAN's, an INTEL SYS 310 computer was selected which allows the parallel processing of such tasks, each appropriate task being allocated to a separate single board computer [6]. The operating system for this computer is Intel RMX86, a real-time multitasking executive. One major feature of relevance to the simulation system is the facility to create tasks using operating system calls. A task can be created, suspended, resumed, put to sleep, deleted and can wait for semaphores or messages. Each program module which simulates the action of a set of entities of a particular type, e.g. buffers, is configured as a task.

At the present state of the work only two single board computers have been used in multiprocessing mode. Ideally if all the tasks are balanced such a multiprocessor system would be 100% efficient, that is, the speed-up of execution of the program would be equal to the number of processors used. Results using just two boards show an initial increase in speed of execution of a simulation model of approximately 1.4. When further boards have been added it is expected that this will improve further, thus providing adequate response to the system designer.

Appendix I lists the steps involved in using the system with data provided by Normalair Garrett Ltd., one of the project's industrial collaborators.

8. CONCLUSIONS

This paper presents a progress report on design tools being developed for assessment of manufacturing systems at Loughborough University of Technology.

Two major areas of specification and design have been recognised. Firstly, an evaluation phase which allows the rapid appraisal of system performance using average measures based on a closed queueing theory model.

Secondly, a detailed emulation phase which generates detailed dynamic information which allows fine tuning decisions to be made concerning system configuration and operation. A high level of detail has been achieved with a reasonable system response.

In the second, emulation phase, it has been shown possible to decompose the total model into a series of modules which can be processed in parallel. Development of this technique on an INTEL SYS 310 micro-computer has shown a considerable improvement in response using parallel processing of the modules on two computers when compared with single board processing.

9. ACKNOWLEDGEMENTS

This work is carried out under an SERC collaborative research grant, reference number GR/B/98128. We acknowledge the support and influence of our industrial collaborators Baker Perkins Ltd and Normalair-Garrett Ltd, in particular Mr T Stratton and Mr R Freeman. The support of all the other members of the FMS Group at Loughborough University and Mr G P Charles is also gratefully acknowledged.

10. REFERENCES

(1) KAY, J.M. and RATHMILL, K. "When Modelling Can Help Your FMS Plans". The FMS Magazine, Vol.1, No. 4 pp 239-244(July 1983)

(2) SOLBERG, J. J. "Mathematical Design Tools For Integrated Production Systems". MTDR No 23, pp 175-187 (September 1982).

(3) DRESCH, K. H. and MALSTROM, E. M. "Physical Simulator Analysis Performance Of Flexible Manufacturing System". Industrial Engineer, pp 66-75 (June 1985).

(4) BONNEY, M. C. "Evaluation And Use Of A Graphical Robot Simulator". Proceedings of 1st International Conference on Simulation in Manufacturing pp 325-330 (March 1985).

(5) LENZ, J. E. "Mast: A Simulation As Advanced As The FMS At Studies". Proceedings of 1st International Conference On Simulation in Manufacturing, pp 313-324 (March 1985).

(6) ROBERTS, E. A. and SHIRES, N. "The Application Of Multiprocessing To Simulation". Proceedings of 1st International Conference On Simulation in Manufacturing pp 85-86 (March 1985).

(7) BUZACOTT, J. A. and SHANTHIKAMER, J. G. "Models For Understanding Flexible Manufacturing Systems". AIIE Transactions pp 339-349 (December 1980).

APPENDIX I: Example Of Use Of Design System – Normalair Garrett FMS

A description of a proposed flexible manufacturing cell from Normalair Garrett is used to illustrate the steps involved in developing a specification, using the interactive FMS software; automatically solving the mathematical model; and automatically configuring and running the emulator. The cell layout is shown in Figure 3.

1.1 Initialisation Of System And Of Specification Build (Figure 4)

Step 1   Start up the interactive FMS software. The initial start up screen, screen 1, will be seen at the VDU.

Step 2   Continue to the main menu screen 2, using key F1.

Step 3   Add information on the elements of the manufacturing system to be built to the database, by selecting the data input module key F1. Once the data has been added return to the main menu.

Step 4   Select the specification build option, key F2 to create the NGL FMS specification.

Step 5   Enter the project name for the specification, screen 3, and select new design option, key F1.

Step 6   Enter the information on screen 4 to identify the specification.

1.2 Interactive Specification Build (Figure 5)

Step 7   Select the machining stations from the database. Screen 5 shows the machining station defined for this case study. Two options are available viz. by specified machine identity (if known) or by interactive specification of machining station parameters.

Step 8   Interactively select the parts to be manufactured. Screens 6 and 7 show selection parameters and screen 8 typical parts data.

Step 9   Select the part buffer stations, see screen 9

Step 10.   Define the material handling systems, see screen 10.

Step 11   Specification build is now complete. Return to the main menu via key F6 and then exit from interactive FMS software, via key F5.

1.3 Configuration and Running of Mathematical Model (Figure 6)

Step 12   Automatically configure the CAN-Q, mathematical model input data by running the program mode. The CAN-Q main menu screen, screen 11, will be seen at the VDU.

Step 13   Select the input data option to input the specification data created with the interactive FMS software, see screen 12.

Step 14   Enter the specification filename, model data filename and material handling average transport time. Then press key F2 to run the model and screen 13 will be seen at the VDU.

Step 15   Screen 14 shows main output menu, by which different soft key options permit outputs to be displayed on relative utilisations and average utilisations of the stations, and overall performance characteristics. Examples of the utilisation figures are shown on screens 15 and 16.

Step 16   If the results from the mathematical model are unsatisfactory re-run the FMS interactive software, and modify the elements within the specification build module to produce a new specification. The user may then proceed from step 12.

Step 17   Once the outputs from the model are satisfactory, exit from the mathematical modelling stage via key F8.

1.4 Automatic Configuration of Emulator

Step 18   Run the software for the automatic configuration of the emulator data for machines, buffers, stations and automatic guided vehicles.

1.5 Layout and Animation Data Creation

Step 19   Draw a layout of the proposed system using the DOGS CAD system.

Step 20   Create an IGES file of the layout.

Step 21   Convert the layout file for output to the selected terminal, in this case an IMLAC vector refresh terminal.

Step 22   Define the automatic guided vehicle (agv) routes and pallet movements using the program interactive software.

Step 23   Enter details on the agv and pallet positions, animated graphic segments and buffers using interactive software.

Step 24   Run a program to create proximity data, which is used to schedule agv movements.

The necessary data to configure the emulation is now complete.

1.6 Configure and Run Emulator

Step 25   Automatically configure the emulation software for this particular specification by running the emulation configuration programs, which access the previously produced data and system description files. At this stage a fully compiled emulation of the proposed manufacturing system has been produced capable of being run in multiprocessing mode on the parallel computers on the INTEL SYS 310.

Step 26   Run the emulation. Files are produced of system performance data and of graphical information for animation purposes.

Step 27   Display the animated output and dynamic utilisation values. (Figures 3 and 7)

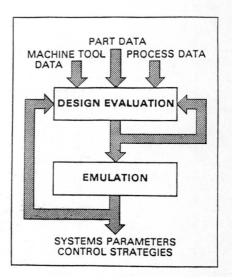

Fig 1   Manufacturing system design

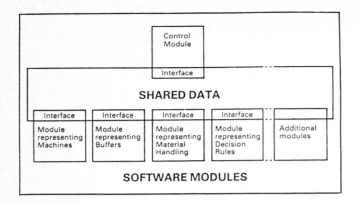

Fig 2   Emulator software implementation

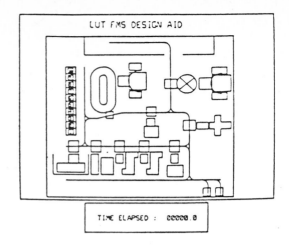

Fig 3   LUT-FMS design aid

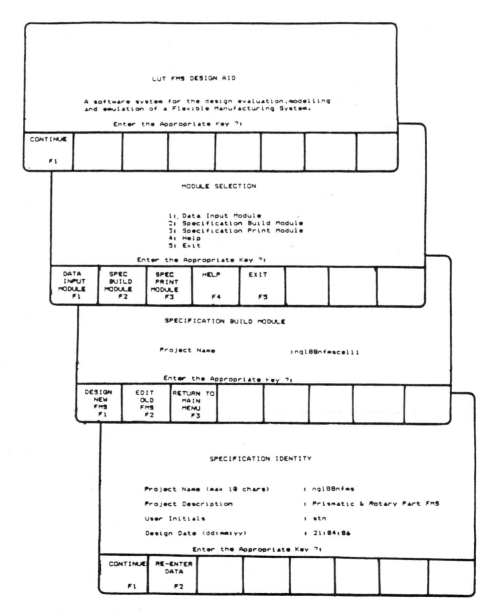

Fig 4   Screen showing initialization of system and specification build

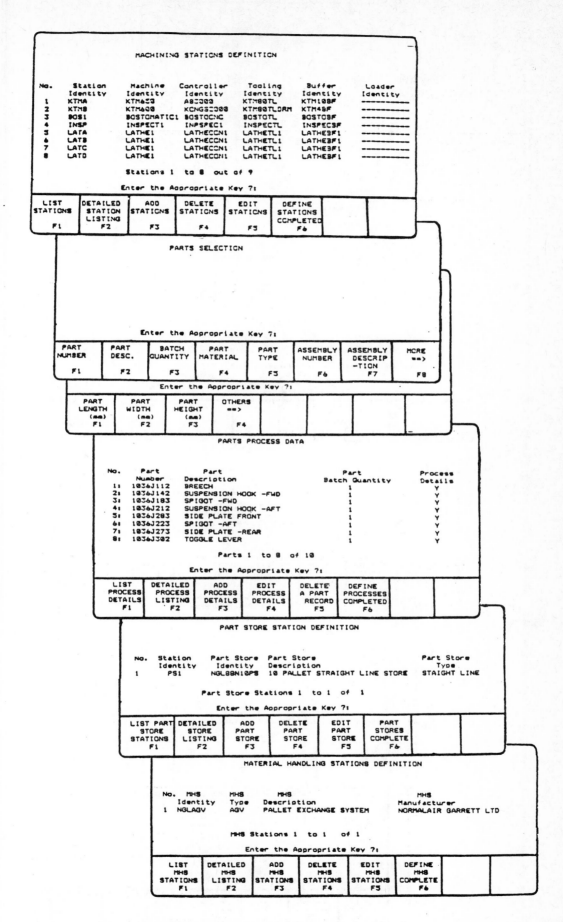

Fig 5   Specification build screens

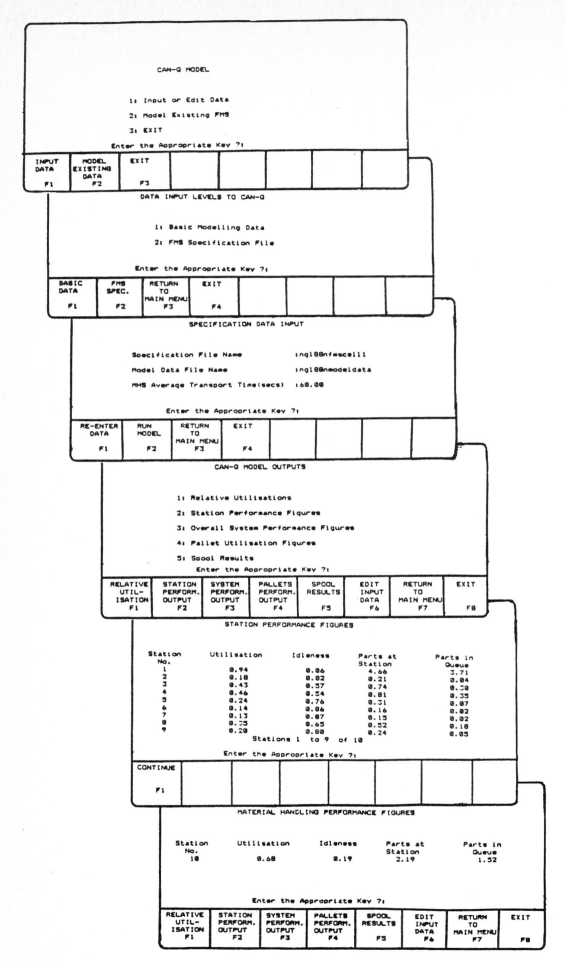

Fig 6   CAN-Q modelling screens

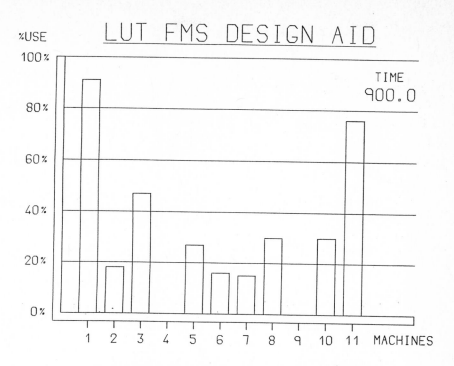

Fig 7   Example of animated histogram output from emulation phase

C366/86

Application of simulation in planning the design of flexible assembly systems

K CHAHARBAGHI, MSc(Eng) and **B L DAVIES**, MPhil, CEng, MIMechE
Department of Mechanical Engineering, Imperial College of Science and Technology, London
H RAHNEJAT, BSc, MSc, PhD, DIC, AMIMechE
Department of Mechanical and Production Engineering, South Bank Polytechnic, London

SYNOPSIS   The purpose of this paper is to demonstrate the use of simulation in planning the design of Flexible Assembly Systems. Industrial case studies are analysed in which alternative strategies are considered in the design of a system. The simulation language which has been chosen is particularly suited to flexible automation. The buffer store size and strategy for layout of the system is discussed, together with the implications of using alternative gripper designs.

1   INTRODUCTION

Flexible assembly systems (FAS) are used to convert components into products, particularly where it is essential to quickly respond to frequently changing market demands. (1) The term "flexible" implies not just a variation in the size, shape and quantity of a product, but also the range of resources needed to produce the different assemblies. The sequence of processes and the mix of machines, material handling devices and tools may vary widely from product to product. FAS are mostly used where a medium variety of product is required in low-to-medium volumes. They enable a number of different products to be manufactured con-currently. Lead times are substantially reduced by closely defining the control and scheduling strategies. Introducing FAS generally involves an intensive capital investment program for which the maximum return on investment is required. This can best be achieved by a close examination at the planning stage of all the options for the configuration, so that the optimal overall efficiency can be realised.

Typically, the layout of such systems consists of a number of workstations, input feeding and orienting devices, and a suitable material handling system for high speed transfer of components. The design of workstations is dependent on the type of product, the selection of the part assembly and the method of feeding. The detailed definition of sub-assemblies requires a subjective approach. This is also required when identifying what degree of flexibility is appropriate to manufacture the whole range of products.

Robot systems for flexible assembly can range from the simple to the complex. The use of "pick-and-place" robots in high volume, fixed automation, assembly can be justified where the system is very structured (2). In such environments, robots with only primitive sensory feedback can be employed to handle a very limited range of similar components. On the other hand, for unstructured assembly systems, with both job variation and low-to-medium volume production, a flexible approach is desired. Within a flexible environment the manipulation of workpieces and tools, for both material handling and assembly, requires a number of basic operations. These can include grasping, holding, orienting, aligning and inserting. In these cases, if a robot is to replace human operators and/or automatic machines, programmable sensing devices are needed to enhance the robot interpretive abilities (3). At an even more sophisticated level, automatic adjustable workstations can be designed to incorporate modular work holders and fixtures as part of the robotic control system. Whether introducing automatic adjustable work-stations or extending the capabilities of current primitive "pick-and-place" industrial robots, an intensive capital investment is required which must be justified by seeing that the system is cost-effective before installation. This is an area where discrete-change simulation is increasingly used to model alternative system configurations at the preliminary design stage as an aid in studying the operational feasibil-ity. A discrete-change simulation method is the experimental study of complex systems, for which analytical techniques often offer no feasible solutions (4). This technique involves con-struction of models that represent and behave in the same way as systems in real life. Once constructed, the models are then put in motion to investigate their behaviour over a desired period of time. In a discrete-change simulation model, changes in the system's characteristics occur at discrete moments in time.

An industrial problem is outlined below in which a number of alternative configurations could be used for the design of a robot based flexible assembly system. The discrete-change simulation technique is employed to enable these various scenarios to be investigated.

2   AN INDUSTRIAL PROBLEM

2.1 The problem definition and the proposed remedial action

A leading engineering firm which manufactures precision products is proposing to redesign parts of its assembly shop to incorporate small flexible assembly units. It is envisaged that each unit will process a family of similar tasks.

A recent production policy appraisal has highlighted the company's inability to respond effectively to changes in consumer demand as well as indicating a decrease in productivity levels. A detailed study of the production schedules has shown protracted lead times which can be attributed to operations related assembly. These findings are not altogether surprising since the assembly tasks account for an average of 54 per cent of the total processing requirements of the company's various products. Presently much of the assembly work is carried out manually or by machines which have to be reset between batch runs of various part types. The excessive set-up times between successive assembly runs have been found to be particularly troublesome accounting for the protracted lead times. The repetitive nature of manual jobs has also been found to result in imprecise and low productivity operations. These major problem areas have forced the company to consider not only a flexible production approach but also the employment of up-to-date technology in its manufacturing resources.

To remedy the above problem areas, the company has decided to design and implement a flexible assembly cell to accomplish the assembly of three different types of products (Here referred to as P1, P2 and P3).

An unassembled component will be located and clamped on a pallet before it enters the cell. The inter-arrival rate of the incoming pallets has been estimated to be 42 minutes per pallet, with an equal chance of any type. A preliminary study of pallet quality has revealed that faulty and incorrectly oriented pallets represent 7.5 per cent of the total input to the cell. Three different configurations have been proposed for the design of this cell.

Figure 1 indicates the first proposed configuration for the cell. This configuration consists of a pallet input location to store any incoming pallet which must wait for its first assembly operation; 3 adjustable machines (here referred to as M1, M2 and M3) to perform the assembly operation; a pick-and-place robot to transfer the pallets between different locations within the cell and a pallet output location to store the assembled pallets. Three men have to be employed if the company adopts this configuration for the cell. The first man is responsible for bringing the unassembled pallets to the cell as well as undertaking an inspection routine on the incoming pallets. This ensures the pallet is correctly presented before the robot transfers it to the machine which performs the first assembly operation. Because of this manual inspection routine, only 5 per cent of the incoming pallets (which are faulty) will be rejected to the scrap bin provided. The company has estimated that this inspection routine will average six minutes per pallet. The other two men will be delegated the responsibility of not only adjusting the machines set up to suit a particular pallet but also emptying the finished goods store. The processing requirements of all types are listed in Table 1. The time taken for the manual adjustment has been calculated on an average basis and is included in the processing times indicated in this table. The pick-and-place robot is a 5-axis cylindrical servo-

actuated robot with elementary force feedback sensing. The robot is programmed as a dedicated manipulator with a fixed gripping force. The end-effector is capable of handling pallets of all types. The transfer time per pallet has been estimated to total approximately 72 seconds. It has been decided to deal with the incoming pallets on a 'first come first served' basis. The in-process competing pallets are transported according to the 'least number of remaining operations' rule. This first configuration of the cell is layed out radially. The collected reliability data has indicated that the time to the next failure of the adjustable machines and the robot is governed by a weibull distribution. The shape and scale parameters for these distributions have been estimated to be 0.995 and 275 hours for the adjustable machines and 0.975 and 350 hours for the robot.

Alternatively, a second configuration, as shown in figure 2, may be employed. Here a 6-axis third generation robotic system with a combination of sensing devices is used to emulate the processing functions of all the adjustable machines, as well as taking on the duties assigned to the pick-and-place robot. The entire assembly work per pallet of any type is performed on a workbench which contains the necessary job-holding fixtures. Table 2 lists the processing durations for pallets of different types. The cell layout is still radial and the inspection routine at the pallet input location is carried out by the robot itself. However, in this case the robot is incapable of re-orientating the incorrectly presented pallets so that both the faulty and the incorrectly oriented pallets are rejected to the scrap bin provided. The inspection routine has been estimated to average 3 minutes per pallet. The robot uses two types of grippers in conjunction with this configuration. A heavy duty gripper with an elementary force feedback sensor is employed for pallet transfer between pallet input location and the workbench, and from the latter to the pallet output location. Two light-duty sensitive grippers are also utilised for inspection of incoming pallets at the pallet input location and for stable grasping, alignment of components and their fitting or insertion at the workbench. The adoption of this configuration would therefore necessitate a gripper change between various robot tasks. The cell will be provided with a gripper store where the robot changes a current gripper to suit its next scheduled task. A change of gripper has been estimated to take 3 minutes. The transportation times of the robot between its successive stops have also been estimated as:
(i)   36 seconds with a light duty gripper and
(ii)   36 or 45 seconds with an empty or a full heavy duty gripper, respectively.
An integrated computer control system will be made responsible for monitoring the operations of the robot within this cell. However this necessitates extensive cell programming to ensure an adequate degree of integration and for this purpose a day-shift programmer must be employed. For this cell only one man will be employed to provide the cell with the unassembled pallets as well as to take care of the assembled ones. It is expected that the proposed robotic system for this cell will be less reliable than the one in the former configuration. This is because of the complexity of the light duty

grippers and the associated sensor controlling software/hardware peripherals. Nevertheless, the predicted reliability data have indicated that the time to the next failure of the proposed robotic system is governed by a weibull distribution whose shape and scale parameters are 0.975 and 250 hours respectively. For this configuration the incoming pallets will be scheduled according to the 'first come first served' rule.

A third configuration has also been proposed for the cell under consideration (see figure 3). In this configuration part handling between the pallet input location and the workbench is carried out by a conveyor. The workbench is allowed to move horizontally, for a specified length, along the conveyor to load an unassembled pallet and/or towards the pallet output location to unload an assembled pallet. A device will also be employed to trigger 'on' or 'off' the conveyor system to ensure the transfer of an unassembled pallet on to the workbench at the right moment in time. The transferred pallet will be held tightly by a pneumatically controlled fixture mounted on the workbench. The transfer time per unassembled pallet has been estimated to average 63 seconds. For this cell the same robot as in the second configuration will be employed. However, the robot tasks are only confined to a series of manoeuvres within the workbench area. The robot only utilises a light duty gripper to carry out the assembly operations on the pallet located on the workbench. Therefore the required processing time of pallets are the same as in the second configuration (see Table 2). It will take the unloading mechanism of the workbench 36 seconds to transfer an assembled pallet to the pallet output location. No input buffer is required. Instead the length of the conveyor is allocated to the queuing pallets. As in the second configuration, a man will be employed to provide the pallet input location with unassembled pallets and also empty the pallet output location from the assembled pallets. With this third configuration the incoming pallets must be processed on a 'first come first served' basis. The inspection routine is carried out prior to queuing at the pallet input location. For this purpose a vision system will be provided which incorporates two TV cameras and a light source placed underneath the pallet input location. Objects are defined by their silhouettes and compared against standard references by a fast processor. A small pneumatic 3-axis manipulator is integrated with the vision system to effect rejection of faulty pallets into a bin provided or effect reorientation of the incorrectly presented pallets. The processing duration for this 'hand-eye' system has been estimated to average 1.5 minutes per pallet. The reliability prediction data have shown that the time to the next failure of this 'hand-eye' system can be considered to obey a weibull distribution whose shape and scale parameters are 0.975 and 300 hours respectively. The conveyor and the loading/unloading mechanism of the workbench are very reliable and can be considered as inherently reliable. The robot failure characteristics can be considered to be the same as in the second configuration. An integrated computer control system monitors all the operations of the equipment within the cell. A day-shift program-

mer will be employed to carry out the cell programming requirements.

The company anticipates the possibility of in-process pallet rejection for all the proposed design configurations. This takes place due to the occurence of equipment failure while performing the assembly task on a pallet. The company is expecting that the time taken to repair any failed equipment within the cell of any proposed configuration will be governed by a lognormal distribution whose parameters (μ,σ) are (2.50, 1.01) (any resulting random sample obtained from this distribution will be in hours). Regardless of what configuration is adopted, the implemented cell will be working around the clock.

Computer simulation is the proper tool which can aid the company in its decision making process to select one of the three proposed configurations and also to prescribe a size for the buffer stock store of unassembled pallets.

2.2 Construction of the discrete-change simulation models

DSSL (5) has been used to construct the models for the purpose of experiments to be undertaken with respect to the alternative configurations outlined above. This simulation language, whose base language is FORTRAN, is particularly suited for this kind of analysis because it possesses specific modules that fully define the essential flexible manufacturing features and allow system reliability characteristics to be included in the simulation models. This language also possesses facilities for obtaining several streams of pseudo-random numbers and uses them to extract random variates from common distributions. The pseudo-random number generation ability allows regeneration of random number streams through the use of specific seeds, thereby enabling the analysts to repeat the simulation experiment by changing an exogenous variable for the system whilst keeping all other variables constant. This ensures a consistant sensitivity analysis with respect to the exogenous variables of interest. Furthermore, a constructed model written in DSSL is very adaptive since it can be readily transformed into another model representing an alternative system configuration simply by adding or deleting a few DSSL statements or changing the input parameters of the existing ones. This accelerates the task of modelling the system reconfigurations.

For discrete-change simulation modelling, DSSL classes entities or objects of a system into three distinct groups: transformable entities, transformers and transformable transformers. Transformable entities are considered as those objects of a system which are transformed from one state or property into another. For example in the design configurations being studied, pallets are classed as transformable entities since their positional and shape properties change as they go through their assembly cell. Transformers are the objects that act as transformation agents. They become unavailable as they engage themselves in the transformation processes. The 'pick-and-place' robot and the adjustable machines in the first design configuration and also the 'hand-eye' system, conveyor and robot in the third design configuration are all examples of transformers. Their manufacturing activities may proceed subject to their availability. Transformable transformers, while being the transformation agents, are themselves transformed in the process so that if they are to proceed to their next scheduled transformation process, they have to revert to the pretransformation state that is associated with this process. The robot of the second design configuration is considered as a transformable transformer since it changes its property as it performs its manufacturing activities and to be able to perform its next scheduled task it has to change its existing property (e.g. once this robot places an unassembled pallet on the workbench with its heavy duty gripper, it must change the heavy duty gripper to the assembly gripper via the gripper store and return to the workbench to perform the assembly task). Since sets are any collection of entities, DSSL applies the above entity classification to the concept of sets.

With most simulation languages, claimed to suit manufacturing systems such as GPSL (6) and SIMAN (7), the creation of an organisational logic structure which describes the operational logic of the simulation models has been delegated to the users. This often results in a laborious task that accounts for a large proportion of model building effort, especially when complex systems are dealt with. On the other hand the ones that are equipped with an associated organisational logic structure, such as SIMON (8) which possesses a 'three-phase' structure, lack some of the flexible manufacturing features and as they stand cannot include system reliability characteristics in the simulation models and thus require modification effort to suit the particular needs (9). Although DSSL is not bound to any organisational logic structures, a robust and coherent structure (see Figure 4) is formulated in conjunction with the use of this simulator which can be utilised advantageously to readily build models of complex systems. The available DSSL modules allow the user to easily construct all the phases of this logical structure.

In the simulation models of the configurations being studied, random number streams have been applied to determine:
(1) The pallet type as it enters the cell,
(2) whether a pallet should be accepted and/or rejected upon inspection,
(3) the time to next equipment failures and
(4) the failed equipment repair durations.
For (1) and (2) in the above list the same two random number streams have been applied in all the simulation models via the application of two specified seeds. This ensures the same pattern of input pallets per type and post-inspection pallet rejections for all the configurations. The following assumptions concerning the operation of the cells being studied have been included in the design of simulation models:
1. No equipment can deal with more than one pallet at a time.
2. There are no other unreliability sources affecting the cells production capability apart from those stated in the systems' description.
3. If a transfer device breaks down while transporting a pallet, this semi-transported pallet will be taken to the intended destination as soon as the device becomes

operable again. If, however, an inspection device breaks down while checking a pallet, this pallet will be inspected by the repairman before he starts the repair operation.

All the constructed models have been simulated for a total period of 26 weeks.

2.3 The analysis of the design configurations on a simulation basis

2.3.1 Discussion of results

Figures 5, 6 and 7 show the variation of inventory level with time obtained from the simulation exercises of the first, second and third design configurations respectively. The simulated operation characteristics of the equipment for each configuration is also shown in the above mentioned figures where operational and non-operational modes are represented by logical digits 1 and Ø respectively. Tables 3, 4 and 5 indicate the results obtained corresponding to the performance of the first, second and third design configurations respectively.

Figure 5 demonstrates that if the cell of the first design configuration possessed no sources of unreliability very little inventory pile-up would be expected for very short periods at the pallet input location. In this design configuration any break down in resource functions will inevitably result in system production incapability. The rising slopes in figure 5 correspond to rapid rates of inventory accumulation as a result of recurring system failures. These failures take place due to impairment of the 'pick-and-place' robot or any one of the three adjustable machines and/or any combinations of these. Figure 5 also demonstrates that the response of the system is rapid in dealing with the rising inventory level when the cause of failure is rectified. The reason for these rapid rates of inventory depletion is due to the under-utilisation of the machines (see Table 3). One way of reducing the high levels of inventory accumulation due to system failures is to introduce resource flexibility in the form of parallel processors at key serving bottlenecks. However, since intermittent system failures are attributed to all the employed resources the effect of such action is likely to be further reductions in utilisations of machines and the robot. This appears to further jeopardise the economic viability of the design with little or no chance of effective payback.

Figure 6 shows that with the second design configuration an ever rising inventory level is to be expected at the pallet input location. System failures take place due to the impairment of the robot. Figure 6 also shows that system failures further exacerbate the rate of pallet accumulation at the pallet input location and the system cannot respond to the rising inventory once a failure is rectified. As compared with the first design configuration, a sharp decline is predicted in the cell production capacity (see Tables 3 and 4). This is partly due to higher pallet rejection rate upon inspection. The simulation results of this configuration reveal a number of weaknesses in the design. These weaknesses are mainly associated with the extent of non-productive activities that are associated with the 'over-

worked' robot. These take the following forms:-
i)   extensive movements of the manipulator with nothing in the gripper and
ii)   frequent gripper-change requirements.
Although the utilisation of the robot is attributed to the periods of its operations indicated by inspection, pallet transfer duties and the assembly work, its optimisation only refers to its effective assembly tasks. Therefore, it is poorly optimised regardless of its relatively high utilisation of 69.62 per cent (see Table 4). There should exist no confidence for this design with such low levels of productivity and ever rising inventory. Even the introduction of a parallel serving robot would entail further financial commitment with a possible reduction in the utilisation level of the existing robot and little hope in achieving a reasonable rate of return on investment, not to mention the untold control and data communication problems in obtaining collision free manipulation.

Figure 7 shows the graph of inventory level over time for the third design configuration. This figure shows the variation of inventory level with time for two distinct locations: one at which the input pallets must wait to be inspected by the 'hand-eye' system and the other at which the inspected pallets must wait to be assembled by the robot. In the third design configuration a combined failure of both the robot and the 'hand-eye' system results in the cell total failure where as the failure of either results in the system being in an operable but derated state. The following conclusions can be drawn from figure 7:
1.   During the first 364 hours of the simulation there are no system failures and no rise in inventory. This reveals that no queue will form if the cell of the third design configuration is always fully operational. This is due to the balanced nature of the input/output characteristic of an equivalent reliable system.
2.   When the 'hand-eye' system is operable a robot failure causes an escalation in inventory accumulation of pallets which must wait for their assembly operations.
3.   When the failed robot is repaired, the rate of reduction in inventory of pallets awaiting assembly is gradual. For this reason the frequent failure of the robot has yielded an excessive inventory of pallets awaiting assembly at the end of the simulation.
4.   A failure of the 'hand-eye' system results in the formation of a queue of pallets awaiting inspection. The repercussion of this failure on the operable robot depends on whether a queue of pallets awaiting assembly exists or not. If such a queue is not already in existence, then the 'hand-eye' system failure causes the robot to be starved of parts. On the other hand, if such a queue exists the robot quite quickly deals with those pallets awaiting assembly, since the queue is not being supplemented by parts which have completed inspection. When the 'hand-eye' system is rectified, it quickly inspects the parts in its queue and passes these to the robot station to form a long queue of parts awaiting assembly. The reason for this phenomenon is because of the large difference in the duration of the inspection and the assembly tasks.

The third design configuration has resulted in a lower production capacity than the first design configuration (see Tables 3 and 5). This is due to the presence of the large queue of pallets awaiting assembly at the end of the simulation. As opposed to the second design configuration, the high robot utilisation in the third design configuration can justify the cost of its use. The problem of the pre inspection queue is not critical since the cell operator could be given the task of inspection whilst the failed 'hand-eye' system is being repaired. Undoubtedly the only major problem with this design configuration is the robot weakness in dealing with the pre-assembly queue which is formed as a result of the robot random failures. However, this problem can be tackled by allowing some degree of resource flexibility. For this purpose an additional robot can be employed within a slightly larger assembly area. Since both robots have their own workbench and loading/unloading mechanism, the incursions of robots into each others working envelope is eliminated and thus the synchronised control of all possible collisions is not required. The layout depicted in Figure 8 shows the newly proposed fourth design configuration.

The extra simulation exercise of the fourth design configuration yields the inventory characteristics shown in Figure 9 and results in the cell performance shown in Table 6. By studying Figure 9, similar characteristics to those exhibited by the third design configuration can be observed with the exception of substantial reductions in inventory levels of pallets awaiting assembly. As in the third design configuration, disturbances to the operation of the 'hand-eye' system results in a pre-inspection queue forming. Subsequently, as the 'hand-eye' system becomes available, the pallets are inspected and transferred to the starved robots for their assembly operation. The rate at which these awaiting pallets are inspected is much faster than the rate at which they are assembled by the duplex-robotised assembly system. This is the reason for a pre-assembly queue which forms subsequent to the repair operation of the 'hand-eye' system. However, as demonstrated by figure 9 and unlike the third design configuration, the system response in dealing with the pre-assembly queue is extremely fast. The phenomenon of the pre-inspection and pre-assembly queue formation can be avoided if the cell operator manually inspects the incoming pallets whilst the 'hand-eye' system is inoperable. A comparison of Tables 5 and 6 reveals that although the additional robot enables the cell to achieve a higher production capacity and improves the inventory, it results in a substantial reduction in the utilisation of the existing robot. This is due to the existing robot being less pressurised in terms of assembling the incoming pallets. The comparison of Table 5 and 6 also indicates that the utilisation of the 'hand-eye' system remains unchanged although an increase in the conveyor utilisation can be seen. The reason for a higher utilisation of the conveyor is due to the more efficient inventory management scheme as described above. Although the utilisation levels of the 'hand-eye' system and the conveyor do not look attractive, their use can be cost-justified since they carry out their tasks much faster and more efficiently than the human operator and can operate outside normal working hours.

Provided that the system does not fail and adequate processing facilities are available, the maximum attainable production rate of the cells being studied is independent of layout considerations. The value of this maximum can be computed from:

$$\text{Maximum attainable prodn. rate (product/day)} = \frac{\text{No. of working hrs per day}}{\text{input pallets arrival (hrs/pallet)}} \times \frac{\text{\% acceptance after inspection}}{100}$$

Therefore for the first, third and newly proposed fourth design configurations, assuming around the clock production, and an input pallet arrival interval of 0.7 hours per pallet with 95 per cent chance of acceptance upon inspection, the maximum attainable production rate is 32 products per day. The production rate achieved by the cell for the first and the fourth design configurations, predicted via simulation is 31 products per day. The production rate of these configurations is slightly lower than the desired one due to the in-process pallet rejection during the simulation and the existence of a queue of pallets awaiting assembly at the end of simulation. Therefore, on the basis of the desired production rate the first and fourth design configurations are satisfactory. The reason that the total predicted production out put of the fourth design configuration is slightly lower than the first design configuration is due to the existence of a larger queue of pallets awaiting assembly at the end of simulation.

The results also indicate that the first design configuration has led to a higher rejection of the in-process pallets. This rejection takes place because for all cell designs, if the equipment performing the assembly operation breaks down while assembling a job, this semi-assembled job is considered a reject and must be dispatched for repair. However, for the first design configuration where the pallets require a higher number of assembly operations on different machines, the chance of equipment impairment whilst assembly is in-progress is higher than with the other design configurations. This leads to a higher in-process pallet rejection.

2.3.2 Recommendations resulting from a comparative study

From the results obtained, the higher production rate and the prompt inventory adjustment offered by the first and fourth design configurations means that one of these must be selected for implementation. Because of the competitive nature of the results, it is vital to compare the various qualitative and quantitative aspects of these two design configurations to arrive at a sound decision. The qualitative aspects should encompass the practical implications involved in the implementation of the designs as well as the future opportunities offered by these configurations. Quantitative aspects refer not only to the system overall performance but also to the capital appropriation and the system running costs.

The results indicate that with the first design configuration, queues of considerable size are likely to form as a result of disturbances to the system operation, even though this configuration is capable of subsequently manipulating

these queues promptly. The space requirements for a proper storage of the high levels of inventory revealed by the simulation, seem impracticable. On the other hand, with the fourth design configuration, if the cell operator is delegated the responsibility of inspection while the 'hand-eye' system is down, no queue build-up occurs except in the very low probability case of impairment of the duplex robotised assembly system in which both robots fail simultaneously. Therefore, with the fourth design configuration, little storage provision is required. The first design configuration can be regarded as a hard-automated assembly cell with limited processing flexibility. This is because, in this design of cell, the 'pick-and-place' robot is a dedicated transfer device and the adjustable machines are usually designed to have a specific degree of versatility and generally handle large volume batch runs. Therefore, this design configuration shows little flexibility in dealing with future changes both in product mixes and specification. On the other hand, the robots in the cell of the fourth design configuration can be programmed to deal with new assembly requirements. Its excess capacity, indicated by the low utilisation levels of the robots, facilitates the ability to increase production levels should the input rate of the cell rise as a result of increased demands. In addition, the duplex robotised-assembly system offers flexibility in product routing should this be required in future. Hence, on the basis of the above discussed assessments, it can be concluded that the fourth design configuration is qualitatively the most favoured.

Table 7 shows the capital appropriation for the first and fourth design configurations. This table indicates that if the flexible approach is selected, then the capital cost incurred is just less than two fold. Table 8 lists the projected running costs for the first and fourth configurations over a period of 26 weeks. The presetting requirements include adjustments of those resources which are related to assembly tasks (e.g. tooling devices, fixtures, grippers etc). The cost associated with presetting requirements is given on an average basis for a unit assembly performed by the assembly equipment. This unit cost is higher for the assembly robot due to the more intricate tooling required. The number of presetting unit costs charged for the configurations is equal to the total production output multiplied by the required number of assembly operations which is carried out by the different assembly equipment. The unit maintenance cost is on an hourly average basis. This includes the repair crew cost for corrective maintenance as well as the cost incurred due to holding spare parts. The maintenance time requirements shown in Table 8 are the results predicted by the simulation exercises. Manning levels are a major running cost in both configurations. For the cell of the first design configuration with 52 weeks in a year, no allowance for the day-shift staff, 20 per cent allowance for the mid-shift staff, 50 per cent allowance for the night-shift staff and 3 cell operators for every working shift, the cost factor for a period of 26 weeks is given by:

The first design configuration personnel cost factor $= 3 \times \left[\dfrac{26}{52} \times \left(\dfrac{100}{100} + \dfrac{120}{100} + \dfrac{150}{100} \right) \right] = 5.5$

Similarly, application of the same approach to the fourth design configuration results in the cost factor of 1.85 for the single cell operator required in every working shift and the cost factor of 0.5 for the day shift programmer. Table 8 reveals that for the design of the first configuration the running costs will be just less than double the fourth configuration. The much lower capital and running costs, and also a slightly higher production output, means that the fourth design configuration attains a greater profit margin and enables the realis- ation of a shorter payback period. Since the qualitative and quantitative assessments are unanimously in favour of the fourth design configuration, it is strongly recommended for implementation.

3. CONCLUSIONS

A detailed industrial application has been used to demonstrate the potential of discrete-change simulation as a powerful technique in the design of flexible assembly layouts.

The large number of modules in DSSL permit a more detailed definition of the system being examined than is possible with many other simulation languages. Furthermore for DSSL a well defined logical structure is formulated. Utilisation of this logical structure diminishes the model building effort, since the user does not have to worry about imposing the correct logic on the system. In this way users of DSSL can readily and accurately model various design scenarios for complex systems.

Analysis of the results of simulation experiments reveal the weaknesses and strengths of design. Remedial actions can be taken against these weaknesses so that sound design policies can be made. These tangible results can best be achieved using the discrete-change simulation method.

REFERENCES

(1) OWEN, A.E. Flexible assembly systems. Plenum Press, 1984.

(2) RANKY, P.G. and HO, C.Y. Robot modelling. IFS (Publications) Ltd, 1985.

(3) ROSEN, C.A. and NITZAN, D. Use of sensors in programmable automation. Proceedings of the 4th international conference on robot vision and sensory controls, 1984.

(4) SHANNON, R.E. Systems simulation : the art and science, Prentice-Hall, 1975.

(5) CHAHARBAGHI, K. Introduction to dynamic system simulation language (DSSL). Centre for Robotics and Automated Systems Report No RCR/85/2, Department of Mechanical Engineering, Imperial College of Science and Technology, 1985.

(6) MEJTSKY, G.J. and RAHNEJAT, H. Introducing GPSL: a flexible manufacturing simulator. CAD, 1985, 17, 219-224.

(7) PEGDEN, D. and HAM, I. Simulation of manufacturing systems using SIMAN. Annals of the CIRP, 1985, 31, 365-369.

(8) HILLS, P.R. SIMON - a computer language in ALGOL. In: HOLLINGDALE, S.H. (Ed.), Digital simulation in operational research, English universities press, 1967.

(9) PAK, H.A., MEJTSKY, G.J., RAHNEJAT, H. and HIBBERD, R.D. An assessment of the existing simulation languages suitable for use in manufacturing environment. SERC report 1 - grant ref. no. GR/C 19325, Imperial College of Science and Technology, 1984.

Table 1 Processing requirements of pallets when the company adopts first proposed configuration

Pallet type	Processing sequence	Processing time per pallet (Minutes)		
		M1	M2	M3
P1	1.M1 2.M2 3.M3	21	21	30
P2	1.M1 2.M2 3.M3	27	18	18
P3	1.M1 2.M2 3.M3	15	24	15

Table 2 Processing requirements of pallets when the company adopts second
and/or third proposed configuration

Pallet type	Processing time per pallet (Minutes)
P1	48
P2	42
P3	36

Table 3 Simulation output indicating performance measures of cell when
first design configuration is employed

PRODUCTION RATE (PRODUCT PER DAY)	PRODUCT	PRODUCTION OUTPUT	EQUIPMENT	PERCENTAGE UTILISATION	NUMBER OF IN-PROCESS REJECTED PALLETS
31	P1	1974	M1	46.58	23
	P2	1902	M2	46.64	
			M3	46.73	
	P3	1935	ROBOT	10.66	

Table 4 Simulation output indicating performance measures of cell when
second design configuration is employed

PRODUCTION RATE (PRODUCT PER DAY)	PRODUCT	PRODUCTION OUTPUT	EQUIPMENT	PERCENTAGE UTILISATION	NUMBER OF IN-PROCESS REJECTED PALLETS
21	P1	1314	ROBOT	69.62	14
	P2	1233			
	P3	1288			

Table 5 Simulation output indicating performance measures of cell when
third design configuration is employed

PRODUCTION RATE (PRODUCT PER DAY)	PRODUCT	PRODUCTION OUTPUT	EQUIPMENT	PERCENTAGE UTILISATION	NUMBER OF IN-PROCESS REJECTED PALLETS
	P1	1885	HAND-EYE SYSTEM	3.57	
30	P2	1825	CONVEYOR	2.23	14
	P3	1849	ROBOT	89.17	

Table 6 Simulation output indicating performance measures of cell when
newly proposed fourth design configuration is employed

PRODUCTION RATE (PRODUCT PER DAY)	PRODUCT	PRODUCTION OUTPUT	EQUIPMENT	PERCENTAGE UTILISATION	NUMBER OF IN-PROCESS REJECTED PALLETS
	P1	1958	HAND-EYE SYSTEM	3.57	
31	P2	1889	CONVEYOR	2.32	17
			ROBOT	55.81	
	P3	1915	REDUNDANT ROBOT	36.63	

Table 7 Capital appropriation for configurations being studied

EQUIPMENT	UNIT PRICE (£)	QUANTITY REQUIRED		COST INCURRED (£)	
		FIRST DESIGN CONFIGURATION	FOURTH DESIGN CONFIGURATION	FIRST DESIGN CONFIGURATION	FOURTH DESIGN CONFIGURATION
PALLET INPUT LOCATION DEVICE	5,000	1	1	5,000	5,000
PALLET OUTPUT LOCATION DEVICE	2,500	1	2	2,500	5,000
'PICK-AND-PLACE' 5-AXIS ROBOT	20,000	1	–	20,000	–
HEAVY-DUTY GRIPPER	900	1	–	900	–
ADJUSTABLE ASSEMBLY MACHINE	60,000	3	–	180,000	–
INTER-CHANGABLE ASSEMBLY FIXTURE	5,250	3	–	15,750	–
HAND-EYE SYSTEM	8,750	–	1	–	8,750
ROLLER-CONVEYOR SYSTEM	5,000	–	1	–	5,000
6-AXIS ROBOT	35,000	–	2	–	70,000
LIGHT-DUTY GRIPPER	3,500	–	2	–	7,000
ASSEMBLY FIXTURE PLUS LOAD/UNLOAD MECHANISM	6,500	–	2	–	13,000
CELL COMPUTER	26,000	–	1	–	26,000
MECHANICAL AND ELECTRICALL INSTALLATIONS	–	–	–	10,000	8,500

TOTAL CAPITAL COST (£)	
FIRST DESIGN CONFIGURATION	FOURTH DESIGN CONFIGURATION
234,150	148,250

Table 8 Projected running costs of configuration for a period of 26 weeks

REQUIREMENT	AVERAGE UNIT COST		NUMBER OF UNIT COSTS REQUIRED		COST INCURRED (£)	
	FIRST DESIGN CONFIGURATION	FOURTH DESIGN CONFIGURATION	FIRST DESIGN CONFIGURATION	FOURTH DESIGN CONFIGURATION	FIRST DESIGN CONFIGURATION	FOURTH DESIGN CONFIGURATION
ASSEMBLY EQUIPMENT PRESETTING	0.25 (£ /ASSEMBLY)	0.60 (£ /ASSEMBLY)	3 X 5,811	1 X 5,762	4,358	3,457
PALLET PREPRATION	0.50 (£ /PALLET)	0.50 (£ /PALLET)	5,811	5,762	2,906	2,881
'PICK-AND-PLACE' ROBOT MAINTENANCE	7.50 (£ /HOUR)	–	208	–	1,560	–
ADJUSTABLE MACHINES MAINTENANCE	12.50 (£ /HOUR)	–	964	–	12,050	–
HAND-EYE SYSTEM MAINTENANCE	–	7.50 (£ /HOUR)	–	356	–	2,670
ASSEMBLY ROBOTS MAINTENANCE	–	10.00 (£ /HOUR)	–	528	–	5,280
REPAIR WORK ON IN-PROCESS REJECTED PALLETS	15.00 (£ /PALLET)	15.00 (£ /PALLET)	23	17	345	255
OPERATOR	8,000 (£ /ANNUM)	8,000 (£ /ANNUM)	5.55	1.85	44,400	14,800
PROGRAMMER	–	12,500 (£ /ANNUM)	–	0.50	–	6,250

TOTAL SYSTEM RUNNING COST (£)	
FIRST DESIGN CONFIGURATION	FOURTH DESIGN CONFIGURATION
65,619	35,593

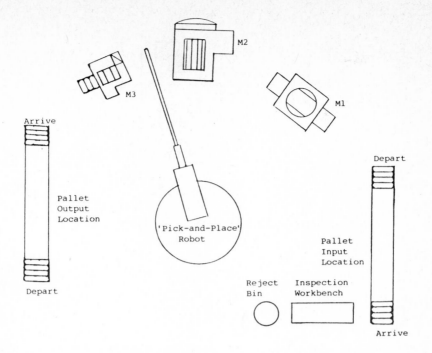

Fig 1   First proposed configuration for cell

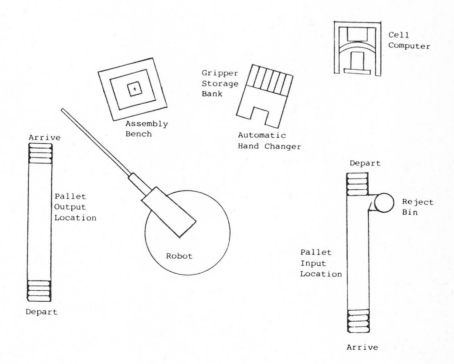

Fig 2   Second proposed configuration for cell

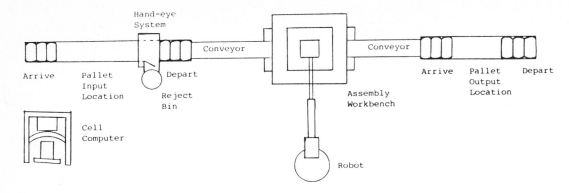

Fig 3   Third proposed configuration for cell

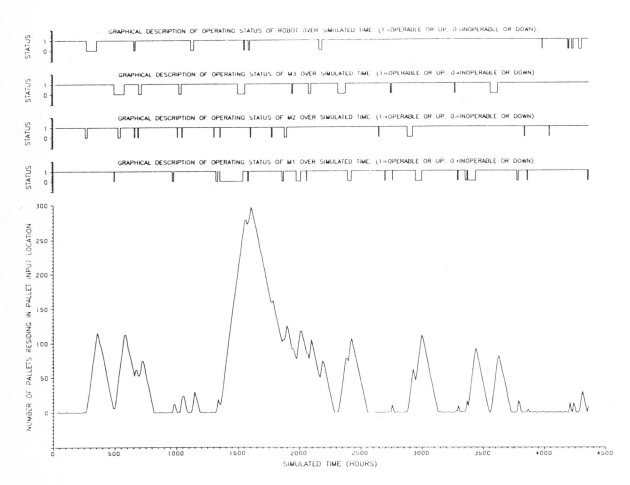

Fig 5   Predicted equipment reliability characteristics and inventory level
over time for first design configuration

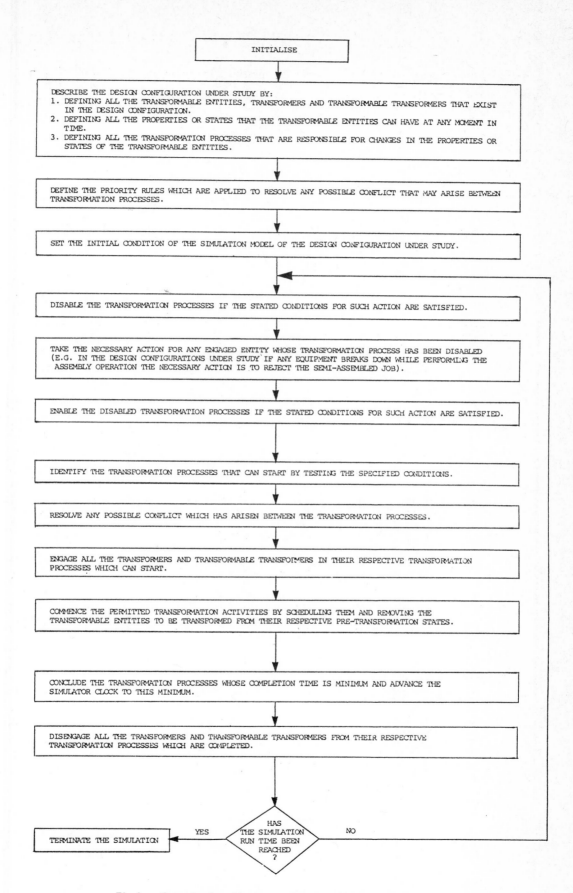

Fig 4   Organizational logic structure for simulation models

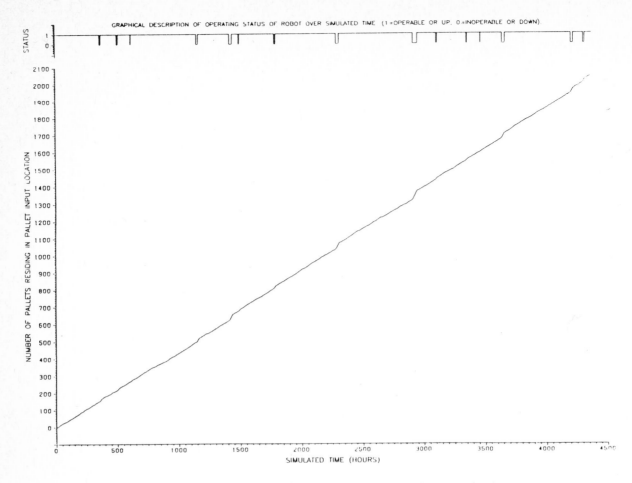

Fig 6 Predicted equipment reliability characteristics and inventory level over time for second design configuration

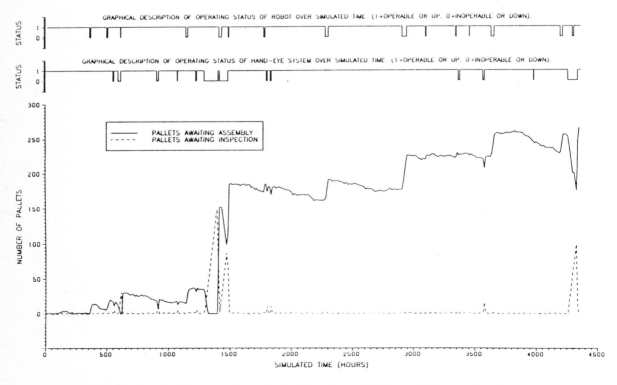

Fig 7 Predicted equipment reliability characteristics and inventory level over time for third design configuration

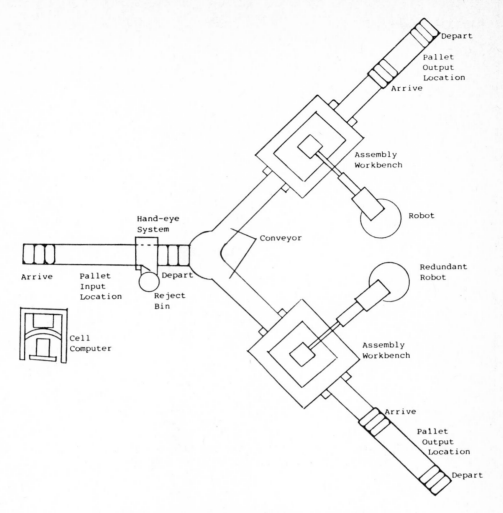

Fig 8   Newly proposed fourth design configuration

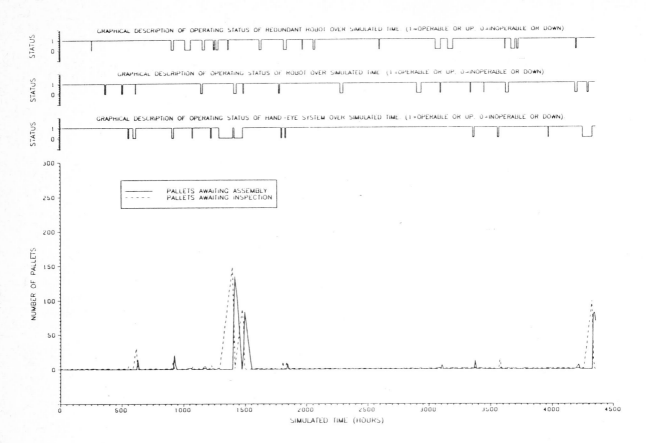

Fig 9   Predicted equipment reliability characteristics and inventory level
over time for newly proposed fourth design configuration

C361/86

Autonomous devices for automated assembly

C J BLAND, BEng(Tech), AMIMechE, P J DRAZAN, Dipl-Ing, MSc, PhD, CEng, MIMechE and S H HOPKINS, BA, MA
Department of Mechanical and Manufacturing Systems Engineering, University of Wales Institute of Science and
Technology, Cardiff

SUMMARY   The next generation of robotic systems will incorporate advanced sensors which will allow
them to compensate for changes in the environment. This increased autonomy is important not only
for robots, but is also a desirable feature for use on conventional assembly machines to compensate
for component misalignments. It is therefore important that the autonomous unit comprises of not
only a sensor, but also contains some means of micromanipulation.

Researchers' efforts to develop autonomous devices for assembly are outlined. In particular the
approach adopted and two of the devices developed at the University of Wales Institute of Science
and Technology (UWIST) are described.

1   THE ASSEMBLY ENVIRONMENT

In many industries assembly still accounts for
a large proportion of direct labour costs and
as a result has been identified as a major
growth area for robotics.

Over the last few decades labour costs have
dramatically increased and there has been the
tendancy to replace manual assembly lines with
purpose built dedicated assembly machines.
However, less than 5% of all manufactured goods
are assembled by such machines. This is
partially due to the machine's high initial
cost and the lack of flexibility and
adaptability compared with manual assembly.

Assembly workers are flexible in the sense that
they can adjust quickly and easily to changes
in the assembly process. Also, human
operators can inspect components for faults
before assembly and can adapt to dimensional
changes in the components. However, manual
assembly tasks are repetitive and tedious
which results in a high labour turnover rate
and inconsistent quality.

In contrast, automatic assembly machines are
custom designed for a particular product and
cannot be easily modified to accommodate
different products or design changes. This
fact in conjunction with their high capital
cost means that these machines are only
economically viable for the mass production
environment and hence have been most
extensively used by the car industry. Also,
assembly machines are very vulnerable to
fluctuations in component quality and if not
carefully controlled, machine jamming can lead
to stoppages in production.

More recently, smaller, more flexible assembly
machines have been developed. These machines
are increasingly being serviced by robots and
are often integrated into manufacturing cells.
Also, with the advent of assembly robots,
flexibility may be further increased. These
machines are suitable for the batch production

environment where the quantity of components
produced is much lower. Hence to maximise
system utilisation the facility for quick and
easy component changeover is important.

For any assembly machine to operate
successfully, misalignment between components
must be prevented or corrected. To prevent
misalignments any jigs and fixtures have to be
precision made for the components being
assembled. This reduces flexibility, since new
jigs and fixtures are required for a product
change, and increases system cost. In
addition, the components themselves must be
maintained within tolerances which are suitable
for the assembly machine. These tolerances
may, therefore, be artificially high for the
components' function and supplementary quality
control on the incoming components may be
required.

Also, there are many industries which cannot
benefit from this technology because the
components they assemble have clearances which
are too small. For example, the assembly of
hydraulic equipment often involves clearances
of 5 microns or less, whereas assembly robots
have typical repeatabilities of 0.02 mm.

To help increase system tolerances and the
application of these systems, research is being
carried out at many institutions to develop
devices which will accommodate misalignment
between components.

2.   PASSIVE ACCOMMODATION

Passive devices are carefully designed
mechanical structures which deflect under
contact forces encountered during assembly so
as to reduce the misalignment. These devices
serve to reduce both the frequency of component
jamming and also the magnitude of the assembly
forces, hence reducing the risk of damage to
components.

The best known device is the Remote Centre
Compliance, developed by the Charles Stark
Draper Laboratory (CSDL) (1). This device is
manufactured under licence and is commercially
available in many countries.

However, there are other forms of passive compliance. One of these is the inherent flexibility in a machine causing it to deflect under applied loads. One recent exploitation of this is the development of the Selective Compliance Assembly Robot Arm (SCARA). This robot was especially designed for extensive use in the assembly environment and has passive compliance incorporated into its structure.

Although passive devices have been successfully used in a variety of applications they have several limitations. Firstly, to reduce assembly forces to a minimum they need a low stiffness and hence are susceptible to low frequency vibrations. Secondly, each unit is designed to suit the component length and therefore has limited flexibility. Finally, these can only be used for cylindrical, chamfered components.

3. ACTIVE ACCOMMODATION

To reduce the limitations of passive devices, active devices have been developed in which the contact forces during assembly are measured by transducers and the signals used to control actuators which compensate for the misalignment.

The simplest devices constructed have been pneumatic systems comprising of low cost back pressure sensors and bellows for actuation (2).

One method of implementing active devices in robotic systems is to attach a sensor onto the robot wrist and use the signals produced to control the robot's movements (3). CSDL developed an instrumented version of their passive unit, the IRCC, consisting of an optical six degree measuring system and passive base. The signals from the device have been used in the feedback loop of a robot controller and enabled chamferless cylindrical components to be assembled (4). A similar device, but having variable stiffness via pressurised spheres and using LVDT transducers has been constructed by Cutkosky and Wright (5).

Six degree of freedom sensors are however, expensive and involve complex signal processing to decouple the axes before the signals can be used to control the robot arm. Therefore, some researchers have reduced the complexity of the sensor by using only three degree of freedom sensors. One example is the sensor developed at Unimation for use on their PUMA robots (6).

Other workers have tried to dispense altogether with any external sensing and instead use the robot's motor currents to measure the misalignment (7).

The disadvantage of all these systems is that it is difficult to compensate for small misalignments because of the high inertia of the robot arm. Also, the clearance of the assembly components must be greater than the resolution of the robot system. Even then there is the need to transfer from the sensor's world co-ordinates to be compatible with the robot's joint configuration.

As outlined earlier, part misalignment is a problem not only for robotics, but also for classical assembly machines. Hence, the inclusion of a micro-manipulating device is an important part of the system.

Systems incorporating separate sensing and actuation have been built, notably by Hitachi Ltd (8) and Brussel (9). The work at UWIST has concentrated on simpler three degree of freedom sensing and two degree of freedom actuation. Particular emphasis is placed on the desire to obtain full information on the insertion process against the need to reduce the complexity of signal processing and hence increase the speed of insertion.

4. ELECTRIC SYSTEMS DEVELOPED AT UWIST

Several sensors have been evaluated and in this report two sensors, one incorporating strain gauges and the other contact switches, will be described and contrasted. Fig 1 shows a system on an IBM 7545 SCARA robot. Although both sensing and actuation are shown on the robot it is equally feasible to separate the two functions, one approach being to mount the actuation beneath the assembly fixture thereby reducing the weight carried by the robot.

4.1 Strain Gauge Sensor

This sensor comprises of a flat plate in the shape of a cross, Fig 2. On each arm of the cross is mounted a strain gauge bridge. The four strain gauge outputs are processed by a computer which monitors the assembly. The X and Y components of the applied force are determined by subtracting the values of the opposite strain gauge bridges. The Z component is obtained by averaging the four values.

Whilst it has the advantage of giving a continuous output it has to be carefully constructed, eg the plate must be flat and the gauges carefully positioned.

4.2 Contact Sensor

Although a continuous output is necessary when a measurement must be recorded, this extravagence is not required when simple force threshold levels are all that is required. This is the case in some assembly tasks which benefit from being monitored by sensors, eg peg-hole assembly. In addition, contact sensors are simple, cheap devices having a high reliability proven in many applications.

One sensor utilising contact switch is illustrated in Fig 3. It consists of a stiffness plate below which is mounted a peg (sensor peg). Above the plate is shown the robot shaft through which the assembly forces are transmitted. Upon application of a force to the gripper the sensor peg moves due to the deflection of the stiffness plate. This movement is detected by the sensors.

In the present design five sensors are used. Four to measure the X and Y components of the applied load and the fifth to measure the axial component of the force. It is envisaged that the four sensors required to measure the X and Y components could be reduced to three.

The advantages of this design are:

1. Stiffness and sensitivity have been separated and can be easily changed.

2. The components do not have to be precision made and the electronic circuitry required is simple.

3. The force components are mechanically decoupled reducing the amount of signal processing.

4. Overload protection is inherent in the design.

5   SENSOR INTERFACE

Both sensors are interfaced to the computer by a Eurocard rack system. The rack allows us to design and build interface modules for a variety of sensors and actuators. The computer is connected to the box via a dual eight bit digital input/output card containing a Peripheral Interface Adaptor (PIA), see Fig 4. One half of the PIA is used to select the interface card; the other half providing eight bits of data transfer to and from the computer.

6   STRATEGIES

Assembly strategies have been developed for chamfered and non-chamfered assembly of cylindrical components (10). In addition work has commenced on strategy development for other more complex assemblies, eg subsea connectors (11).

The programs were intially developed using a combination of BASIC and 6502 machine code. They have now been rewritten in the language FORTH. This stack based language was designed for control applications and has recently seen an up-surge in the number of applications in robotics.

FORTH is a highly efficient language comprising an interpreter, compiler, assembler and multi-tasking capabilities. A program is written as a series of small discrete routines which are given a name and then can be used in subsequent routines. This lends itself to the Top down, Bottom up philosophy of programming. Each routine may be tested on its own leading to rapid program development. Also, because the finished program comprises of a series of machine code calls, memory usage is small.

6.1   Strain Gauge Sensor Strategies

Assuming that there is chamfered contact between peg and hole, a lateral force vector is produced, its direction being towards the centre of the hole. This is measured by the sensor and the X and Y motors are driven in the direction of the vector (10).

If the chamfer on the peg does not make contact with the hole, or if the components are unchamfered, then a search pattern is implemented. The search pattern follows a

spiral where the misalignment is unknown and a wave pattern if the misalignment is always on one side of the hole.

6.2   Contact Sensor Strategies

The contact sensor cannot produce a vector, it only indicates one of eight movement directions for the component misalignment as shown in Fig 5. However, because of the reduced signal processing the speed of insertion is increased.

The assembly algorithm used for the assembly of chamfered cylindrical components is shown in Fig 6. The vertical axis is driven down until assembly is complete or, if there is misalignment, contact is made with one of the switches. The vertical switch (switch 5) is first examined and if ON an error condition is assumed and the assembly aborted. The misalignment zone is determined by examining the switch conditions and when found the micro-manipulation table is moved accordingly. This is repeated until assembly is completed.

Experiments are now proceeding to use this device for the assembly of unchamfered components. In addition, other applications are being examined.

7   CONCLUSIONS

The number of sensors being used by industry in the robotic and assembly environment will only increase dramatically when they are reliable, simple, cheap and fast acting.

It has been the aim of the work at UWIST to develop devices which will provide some intelligence to compensate for the misalignments occuring during assembly. In this report two sensors used as part of an autonomous assembly system have been described and contrasted.

Further work will concentrate on exploiting and furthering the strategies developed. This will be performed in conjunction with a database on industrial assemblies which will be used to extract the common generic features of assembly.

8   REFERENCES

(1) WATSON, P. C. Remote center compliance. US Patent No 4098001, July 1978.

(2) DRAZAN, P. J., HOPKINS, S. H. Semi-autonomous systems for automatic assembly. Annals of CIRP, 1984, 33/1.

(3) PILLER, G. A compact six degree of freedom sensor for assembly robot. Proceedings 12th International Symposium Industrial Robots/6th International Conference Industrial Robot Technology, June 1982, 121-129.

(4) DE FAZIO, T. L., SELTZER, D. S., WHITNEY, D. E. The instrumented remote centre compliance. The Industrial Robot, December 1984, 11, 4, 238-242.

(5) CUTKOSKY, M. R., WRIGHT, P. K. Compliance system for industrial manipulators, US Patent No 4458424, 10 July 1984.

(6) SPALDING, C. K. A three-axis force sensing system for industrial robots. Proceedings 3rd International Conference Assembly Automation, May 1982, 565-576.

(7) LEIGEOS, A., DOMBRE, E., BORREL, P. Learning and control for a compliant computer controlled manipulator. IEEE Transactions Automatic Control, December 1980, AC-25, 6, 1097-1102.

(8) GOTO, T., TAKETASU, K., INOYAMA, T. Control algorithm for precise insert operation robot. IEEE Transactions Systems, Man and Cybernetics, January 1980, SMC-10, 1.

(9) BRUSSEL, H., SIMONS, J. Automatic assembly by active force feedback accommodation. Proceedings 8th International Symposium Industrial Robots, 1978, 181-193.

(10) DRAZAN, P. J., HOPKINS, S. H., BLAND, C. J. Autonomous assembly devices and related control strategies. IEE Control Engineering Series No 28, Robots and Automated Manufacture, 1985, Paper 9.

(11) CHUCAS, P. D. Sensor assisted assembly of a subsea connector. UWIST M Eng Thesis, October 1985.

Fig 1   Assembly system mounted on selective compliance assembly robot arm (SCARA) robot

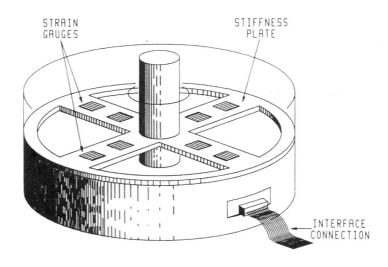

Fig 2   Strain gauge sensor

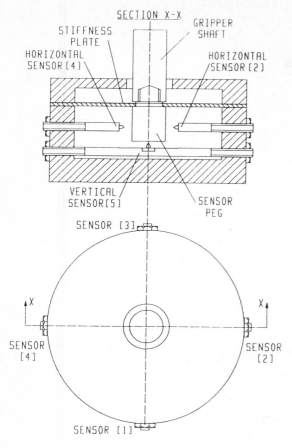

Fig 3   Contact switch sensor

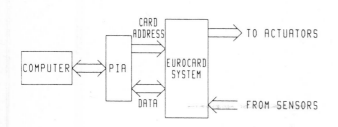

Fig 4   Interface configuration

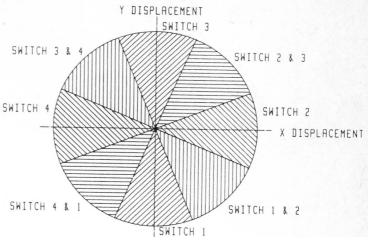

Fig 5   Misalignment zones

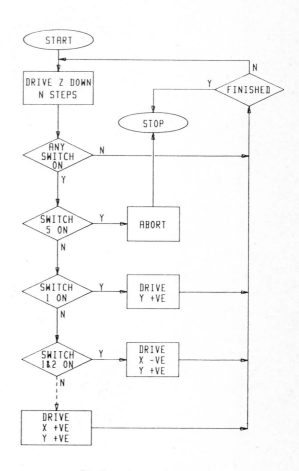

Fig 6   Assembly flow chart

C373/86

Spatial differential energy for the detection and location of multiple moving targets

P D CHUANG, PhD, DIC, **J LAWRENCE** and **C BESANT**, PhD, DIC, CEng, FIMechE
Department of Mechanical Engineering, Imperial College of Science and Technology, London

SYNOPSIS An extremely fast algorithm using differences in the accumulated spatial energy between successive video frames to detect and locate multiple moving targets within the frame is described. This algorithm is modest in its digital memory requirements, and as such, a conventional digital frame store is not necessary. Applications could exploit either the reduced memory requirements, or the speed in locating moving targets, or both.

1 INTRODUCTION

Dynamic scene analysis has recently been attracting increasing attention in computer vision (1-3, 6-10), with applications in a large variety of diverse situations. Examples of such wideranging applications include gathering data on the behavioural patterns of fish, tracking aerial targets in military applications, motion detectors in surveillance, and components tracking in robot vision. In these analyses the locating of moving targets is always the fundamental problem, especially in segmenting the moving target or targets of interest from the background clutter so that further analysis can be performed on the segmented targets.

There are various ways of locating the two dimensional coordinates of targets, all of which depend on the ability to discriminate the targets from the background. Given a single frame, the simplest form of target location is to remove the grey levels that are not associated with the targets of interest, for instance, binarising, thresholding or intensity band pass filtering based on the histogram information. Other classical techniques use template matching (binary image or grey) or location through segmentation by edges (4,5). Unfortunately, these techniques inherently require some *a priori* knowledge of the targets to be located. In addition, the computational requirements for real time analysis may be excessive. This is certainly true if the target is easily confused with the background.

The task of locating targets is simplified if the targets in question are in motion relative to the background. Thus, it is not unusual that most dynamic scene analyses exploits these temporal changes between frames. Even so, the sophistication of these analyses can range from simple differences between successive frames (6,9) to optical flow and motion field analysis (10). Needless to say, the information available from different techniques depends on the complexity of the analyses and the selected technique is usually balanced against the execution speed.

Given more than one frame, the easiest way of locating targets in a scene with a complex background is to compare a reference frame without any targets with one containing targets. Although some forms of differencing techniques have already been previously documented, Jain (1984) formally explains them in his paper (8). A *difference* picture can be prepared by comparing corresponding pixels of the reference frame and the target frame, giving positive difference pictures, negative difference pictures, absolute difference pictures or accumulative difference pictures depending on the comparison operator. Disregarding noise, non-zero pixels would appear only where the target or targets are located. Given this information, multiple target locations and their spatial boundaries can be deduced.

2 OBJECTIVE

The objective of this paper is to outline a fast algorithm, as well as the hardware implementation of the algorithm, in order to extract two dimensional multiple target positional information from a single stationary video camera system.

This paper suggests a modifed difference approach while still retaining a high execution speed and simplicity. It is not the intention of this paper to analyse the target(s) in motion in detail, but to provide a quick indication of the location and spatial boundaries using multiple frames so that appropriate analyses can be carried out within the constrained search area. Thus, this front end technique is extremely useful when utilised in conjunction with vision systems incorporating electronic windowing or similar facilities. As with algorithms using the successive comparison philosophy, the obvious disadvantage is that more than one frame with the same background scene is essential and therefore, the observing camera has to be stationary. If, however, the camera is also in motion, differencing techniques are only applicable if the background is uniform, or is removable by some means (e.g. thresholding), or if the relative motion of the background is negligble in relation to the target. Tests have shown that the algorithm can reliably locate up to three regularly sized targets within the camera's field of view. Locating more than three target locations is still possible although ambiguity may arise.

3   TARGET LOCATION THEORY

3.1   Accumulative spatial energy

To accomodate the data throughput for target location and to overcome the processing problems of manipulating and storing large amounts of pixel data, this paper suggests the use of a reduced set of spatial information, i.e., the use of accumulated spatial energy or intensity values. Initially in the form of row and column summations for locating single targets and incorporating both diagonal summations for multiple targets. Using this reduced data set a full resolution frame store is therefore unnecessary.

In order to locate targets in real time, video information of each frame is compressed into four arrays per frame, i.e. row, column, positive diagonal and negative diagonal. Figure 1   shows an example with a 6 x 6 pixel image. In an image frame of m rows and n columns, accumulating the spatial energies of every pixel in each row will result in m row elements, R_i, with each element containing the total spatial energy across that row. Similarly n column elements, C_j, containing the total spatial energy along the column can also be obtained. Repeating the summation along the diagonals will correspondingly give two diagonal arrays, positive diagonal, $+D_k$, and negative diagonal, $-D_k$. Mathematically,

$$R_i = \sum_{x=i,y=0}^{y=m-1} g_{x,y} \qquad (3.1)$$

$$C_j = \sum_{x=0,y=j}^{x=n-1} g_{x,y} \qquad (3.2)$$

$$+D_k = \sum^{\text{all } k \text{ +ve diagonal pixels}} g_z \qquad (3.3)$$

$$-D_k = \sum^{\text{all } k \text{ -ve diagonal pixels}} g_z \qquad (3.4)$$

where R_i is the total spatial energy of row i such that $0 \le i < m$,

C_i is the total spatial energy of column j such that $0 \le j < n$,

$+Dk$ is the total spatial energy along the positive diagonal and $-Dk$ is the total spatial energy along the negative diagonal such that $0 \le k < m+n-1$,

g is the grey intensity of the pixel,

m,n are the resolution of the rows and columns respectively,

x,y are the row and column addresses respectively, and

z is the corresponding +ve or -ve diagonal address.

If we consider for the moment, a image size of 256 x 256 pixels, the accumulation will result in 256 row elements, 256 column elements, 511 positive diagonal elements and 511 negative diagonal elements per frame. The immediate effect of summation is to reduce 64K pixels (256x256) to about 1.5K elements, giving a reduction ratio of more than 21 to 1 if the summed elements are twice as large (e.g.16bits). The savings will become more apparent when a succession of frames are considered.

3.2   Single target location

To locate targets, two compressed frames are compared. For simplicity, consider a frame of a complex scene without any targets as shown in figure 2a in which the row and column accumulated spatial energies have been found. Taking a second frame of the same scene, similar to the first except for the inclusion of a single target will yield a different set of total spatial energies. Figure 2b shows the cursor and spatial boundary overlaying the located target and its shadow. In a noiseless situation, any difference between the two sets of spatial energies will only be at the row and column locations containing the target. Background clutter and stationary objects in the scene are repeated in the same spatial locations and are reflected in both sets of accumulated spatial energy. Thus, taking corresponding differences of every element between the two sets of parameters, in the form of an absolute difference operation, will only result in a non-zero value at that row and column coordinate at which the target is

118

located. Therefore, the spatial boundaries of the target can be found from the first and last - row and column non-zero difference locations as shown in figure 2b. The location of the target can also be represented by the row and column addresses where a maximum difference occurs.

The use of accumulated spatial energies is by no means new! Commercially available motion detectors operate along the same principle. However, these systems are either constrained to a single target or to predefined zones within the field of view, or they require some form of frame store. It should be noted that the accumulated energies do not refer to the individual pixel values and recovery of the original frame after accumulation is not possible.

3.3 Multiple target location

Consider another frame of the same complex scene with two targets rather than one, figure 2c. The accummulation and difference procedure is again repeated. Comparing the compressed reference frame with the compressed target frame will again result in energy discrepancies at only the row and column locations containing the targets. Again, stationary objects in the image are assumed to be background noise and thus will be transparent to the target location algorithm. However, the same target locations as shown in figure 2d will result in identical discrepancies. There are now four possible target boundaries, two containing 'real' targets and the other two containing 'imaginary' targets.

To overcome this ambiguity problem, this paper proposes the use of two additional sets of accumulated spatial energies, namely, the positive and negative diagonals. To locate the two targets, only the real locations will produce non-zero difference values for all four parameters. As the number of pixels along different diagonals depends on the diagonal, ranging from 1 to 256 pixels for a 256 square image, non-zero differences are scaled in proportion to the number of pixels in that diagonal. This final verification is always correct for two targets providing the targets do not overlap or touch one another. If this algorithm is expanded to include three objects in any arbitrary location, target location will still be correct although the possible target locations before verifying with the diagonals is now a maximum of nine due to the permutation of locations.

4   PRACTICAL PROBLEMS

4.1 Multiple target ambiguity

In situations where more than three targets appear within the field of view, ambiguity may exist. Figure 2e demonstrates an extreme case where the centre target is 'shadowed' by the surrounding targets so that a difference in all four accumulated spatial energies will result even if the centre target does not exist. Figure 2f highlights another example of ambiguous target locations. Ambiguity occurs when one or more spatial energy arrays are computed from pixel energies of more than one target. In the worst case where numerous targets exist, target or targets will be lost. However, in typical cases where three or four regularly sized targets are within the field of view, ambiguity only results in a larger spatial boundary for the targets. Thus ambiguity is dependent on the resolution of the video frame prior to compression as well as on the size of the targets. The larger the target, the greater the probability of ambiguous locations. In the worst case, the spatial boundary will indicate the maximum boundary in which disturbances are taking place.

To partially resolve the problem of ambiguity, further research is being carried out into the use of predictive tracking[10]. This is performed by logging the information of each target across numerous frames up to the frame when ambiguity occurs. This allows the prediction and estimation of target locations when ambiguity exists until ambiguity ceases to exist.

4.2 Noise

Up to now, we have assumed that the images are noise free so that changes in the spatial energies are due to target motion. However with real images, the effect of noise will corrupt the comparison between spatial energies of the reference frame and the current frame, although random pixel noise and quantisation noise is usually integrated by the process of accumulation. To reduce the effects of noise, only absolute differences above a predetermined threshold are accepted for spatial boundary computations. Alternatively, targets can be located by the row and column coordinates where a maximum difference or percentage difference occurs, while still reflecting a change in the diagonal elements at that coordinate. In general, as noise level increases, mislocation will be more frequent. It is for this reason that this paper suggests a maximum frame resolution of 256 x 256 pixels to reduce random noise errors. For a frame resolution of 512 x 512 pixels, either alternate pixels and rows are used, or four segments of 256 x 256 pixels.

To improve target location further, the absolute difference between the two frames can be smoothed by a three element moving average filter before thresholding as previously discribed.

Motion distortion, as a form of noise, due to the long persistence of the scanning system can also corrupt spatial information. Camera tubes (e.g.Vidicon) are normally designed with a persistence longer than 20ms. This persistence phenomena of 'comet tailing' is inherent in scanning system utilising scanning tubes and therefore limits their use in target tracking. To remedy this, solid state cameras (e.g.CCD) which do not have any persistence lag are preferred. Besides the problem of persistence, motion distortion can also be due to the target velocity . This form of motion distortion is difficult to remove and can only be removed by :-

- increasing the scanning rate;
- electronically gating the scanning system;
- increasing the optical field of view; or
- combinations of the above.

Increasing the optical field of view reduces the relative target motion. Unfortunately, besides the difficulty of modifying a camera to achieve a faster scanning rate, high speed scanning requires image intensification. It is for this reason that Gated Silicon Intensified Target (GSIT) CCD cameras are recommended.

4.3 Reference frame selection

As comparisons are taken between the reference frame and successive frames, the choice of the reference frame is important. The reference frame is usually the latest frame prior to the appearance of targets within the field of view. This is true for a stationary camera under controlled lighting. However, due to random changes in the ambient lighting conditions during the appearance of target(s), the reference frame may not be valid after a series of comparisons. Thus the reference frame has to be updated with information from the current frame and comparisons are now made between the new reference frame and the next frame. If the target is totally displaced in the new frame, two target locations will appear, giving the old and new target locations. This is shown by figure 3 where a swinging target held by a transparent string demonstrates the results of partial displacement and total displacement during the next frame. If the displacement is not total, the new and old target locations will also overlap and a larger spatial boundary will result. By keeping record of old target locations new locations can be deduced.

5 HARDWARE

As with most target location algorithms, speed is crucial. The accumulation of spatial energy for array comparisons and spatial boundary computations can easily be performed by purpose built hardware at speeds in excess of the standard scanning rate. Furthermore, as a frame rate faster than the standard video rate is presently beyond the ability of most conventional frame stores, purpose built hardware is recommended for a more cost effective solution.

To appreciate the pixel throughput required for dynamic scene analysis, let us consider a video frame rate of 50Hz (20ms per frame). Using a 512 x 512 pixel frame resolution, this is equivalent to a pixel rate of 10Mpixels per second including blanking periods. If the video rate is doubled or quadrupled, this corresponds to about $20Mpixs^{-1}$ and $40Mpixs^{-1}$ respectively, with each pixel containing grey level information (typically 6bits or 8 bits wide).

The block diagram of the hardware implementation is shown in figure 4. For each input pixel, the accumulation of the four spatial energies is executed in parallel under the direction of the memory controller and address generator. The compressed spatial energies are stored in two groups of memories, one containing the reference (previous) frame and the other containing the target (current) frame. ECL memory devices as well as ECL logic devices are used for a faster cycle time. Target location and spatial boundaries are therefore available at standard or increased frame rate. The other blocks in the figure show the digitisation of the incoming video signal, digital comparators, data buffers, threshold controller, and target location cursor/marker (only at standard video rate).

The hardware implementation in general is only constrained by the speed of commercially available A/D convertors and the scanning speed of video cameras. Using a non-standard GSIT CCD video camera and an A/D conversion rate equivalent to 100MHz (or 10ns per pixel), an execution speed 10 times faster than a standard 512X512 CCD camera, or 20 times at 256X256 resolution is achievable. Alternatively, high speed target location can be achieved by multiplexing among numerous channels.

6 CONCLUSION

Multiple target location is possible using spatial differential energy by comparing the corresponding differences between two video frames. The simplicity of the logic and the reduction in the memory requirements is particularly attractive in applications where a limited number of targets are to be simultaneously located at frame rate. The possible hardware implementation at non-standard frame rates also implies that previously untrackable targets can now be located without excessive costs.

REFERENCES

(1) Huang, T. *Image sequence analysis* Springer-Verlag, Heidelberg, FRG (1981).

(2) Roach, J. W. and Aggarwal, J. K. 'Computer tracking of objects moving in space' *IEEE Trans. Pattern Anal. Mach. Intell.* Vol 1 (1979) pp 127-135.

(3) Roach, J. W. and Aggarwal, J. K. 'Determining the movement of objects from a sequence of images' *IEEE Trans. Pattern Anal. Mach. Intell.* Vol 2 (1980) pp 554-562.

(4) Duda and Hart. *Pattern Classification and Scene Analysis* 1973.

(5) Batchelor, B. G. *Pattern Recognition* 1978.

(6) Jain, R. and Nagel, H-H. 'On the analysis of accumulative difference pictures from image sequences of real world scenes' *IEEE Trans. Pattern Anal. Mach. Intell.* Vol 1 (1979) pp 206 -213.

(7) Jain, R., Martin, W. N. and Aggarwal, J. K. 'Segmentation through the detection of changes due to motion' *Comput. Graphics Image Process.* Vol 11 (1979) pp 13-34.

(8) Jain, R. 'Difference and accumulative difference pictures in dynamic scene analysis' *Image and Vision Computing* Vol 2 No 2 May 1984 pp 99-108.

(9) Horn, B. K. P. *Robot Vision* The MIT Press McGraw-Hill Book Co 1986.

(10) Hunt, A. E. and Sanderson, A. C. 'Vision-Based Predictive Robotic Tracking of a Moving Target' Department of Electrical Engineering and The Robotics Institute, Carnegie-Mellon University, Pittsburgh CMU-RI-TR-82-15 Jan 1982.

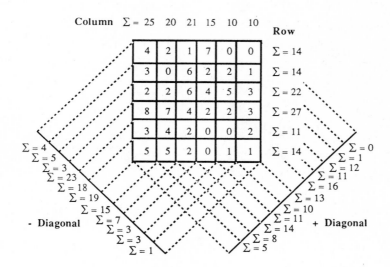

Fig 1   Example of total spatial energy calculation

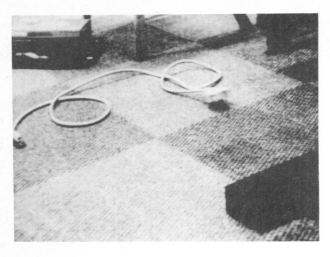

Fig 2a Complex scene without any targets

Fig 2b Located target (and shadow) with cursor and spatial boundary overlay

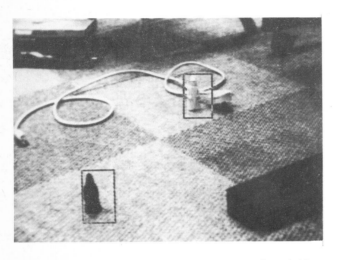

Fig 2c Complex scene with two targets and overlaid boundary

Fig 2d Target discrepancies with coordinates similar to Fig 2c

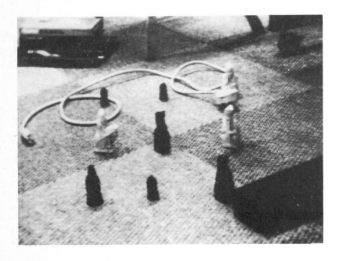

Fig 2e Ambiguity where centre target cannot be confirmed

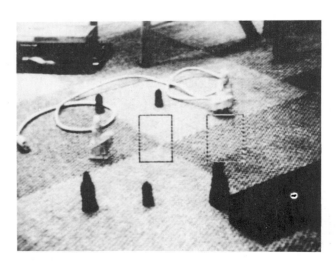

Fig 2f Another example of ambiguous target locations that cannot be confirmed

Fig 3a   Reference frame

Fig 3b   Partial displacement

Fig 3c   Total displacement

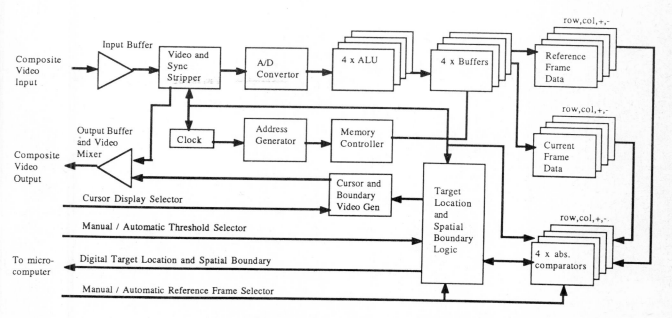

Fig 4   Block diagram of hardware implementation

C363/86

Progress in the design of electric drives for modular robotic systems

G G ROGERS, BTech, **R H WESTON**, PhD, BSc, **R HARRISON**, BTech and **A H BOOTH**, BSc, MSc
Department of Engineering Production, Loughborough University of Technology, Leicestershire

SYNOPSIS   The requirements of electric drives for modular robot systems is considered.   Analysis shows that simplicity of use combined with a variety of move types is one of the major requirements. Secondary information concerning "performance objectives" and "styles" of control are suggested as ways to futher optimise performance capabilities of the individual modules.   A hardware system is proposed and the control strategies examined.

1   INTRODUCTION

Work in the area of modular robotic systems has been a research area of interest to the Department of Engineering Production at Loughborough for a number of years, with initial work being centred on the design of pneumatic motion control systems.   Generally robot modules employing pneumatic motion controllers, of the type evolved at Loughborough, have been found to be particularly suited to the point to point positioning of small to medium payloads where they provide significant cost advantages over their electric and hydraulically actuated counterparts (1).   However, the use of electric actuation offers an ability to accurately define motion profiles and hence extend the application areas in which modular robots are traditionally employed.   Over the last few years a variety of proprietary electric driven single degree of freedom modules have become available from a number of sources.   The introduction of such modules has been promoted by corresponding developments in the design of electric drive system elements, eg

(i)   improvements in switching amplifier design offering high power capabilities with much improved power factors compared to earlier thyristor amplifiers (2),

(ii)   improvements in the availability and performance of dc motor types and forms (3),

(iii)   the availability of a range of proprietary transmission system elements, and

(iv)   the emergence of "first generation" micro-processor based controls suitable for electric motion controls.

With this knowledge a program was undertaken to investigate the use of electric drives in modular robotic systems.   The work reported here concentrates on the design of a single axis module (SAM) and an associated single axis controller (SAC): these elements representing basic building blocks for more complex robotic systems.   Another essential ingredient is a supervisory computer which is responsible for issuing commands to the SAC's so as to produce the desired motions of the robotic system.   Work in this area is also being pursued by researchers within the Department where the particular requirements of programming and operating sytems for concurrently operating multiaxis manipulator groups, demonstrating user defined kinematics and dynamics, are being studied.

The primary function of the SAM and the SAC is to perform the motions as required by the supervisor.   The criteria on the type of motion required will be different depending on the task being performed.   When defining the desired motion a set of motion descriptors must be transmitted to the SAC.   Similarly the desired motion can only be performed provided it is within the capabilities of the SAM.   The direct result of this realization is that each axis controller has a set of basic axis control capabilities which are selectable from the supervisor.   Once the type of motion control has been selected then secondary move parameters sent by the supervisor would specify further requirements.

The following "types" of motion are important in robotic application areas:

(1)   Point to Point

(2)   Point to Point with constant velocity over part of the stroke, ie trapezoidal velocity profile

(3)   Contouring, this being defined as the ability to pass through specified points with defined velocity.   Accuracy both in positional and velocity terms being a specified parameter

Along with the command type can be included another parameter, namely a performance objective.   This parameter defines any secondary criteria that needs to be met in unison with the required motion.   For example, in moving between two points it is possible to achieve the motion in a multitude of ways.   Examples include minimise time taken, never exceed a specified velocity or acceleration, minimise the temperature rise in the motor (ie maximise the duty cycle capability of drive), etc, etc.

In addition to the command type and its specified performance objectives a third parameter can be supplied which defines what might be called "style" - namely the type of control approach used whilst executing the specified command with the specified performance objective. The "styles" thought to be of use in robotic work are:

(1) Robust - predefined gains suitable for the complete range of loadings for a particular module

(2) Adaptive - the controller attempts to improve upon its performance

(3) User defined - knowing parameters of the load system the user can specify appropriate data to the SAC, from which the optimum gains are determined

It is important that the "style" is selectable from the supervisor since often the control approach needs to be different depending on the task at hand. For example, in a production process requiring the constant machining of items, a robust style of control (with fixed gains) would be preferable to an adaptive control strategy which might result in unwanted product variations. Similarly in a pick and place task which moves various different masses the criteria may be to minimise time taken and yet maintain acceptable damping. Here constantly adaptive features which improve upon the time response would be advantageous.

2 ELECTRIC DRIVES

Essential manipulator building blocks in modular robotic systems are rotary and linear drives. Here we will concentrate on the design of linear drive systems for which the two main contenders in choosing a drive transmission are lead screws and rack and pinion drive elements. Belt transmission can be used in gantry type systems where high static accuracy and dynamic performance is not so critical. Rack and pinion systems have tended to be used for high speed applications where some backlash is allowable while lead screws have traditionally been used for slower speed applications. For lead screws, low friction and backlash can be obtained by the use of re-circulating ball nuts and zero backlash can be obtained by using two nuts which are pre-loaded against each other. Recently a number of high lead devices have become available to meet the requirements of the robotic industry. As a result it is now possible to obtain re-circulating ball screws which have leads equal to the diameter of the lead screw shaft. Also, a number of novel lead screw type devices which use small rollers to follow spiralling grooves around a plain shaft are available for even higher lead/shaft diameter ratios (4).

The first linear electric SAM prototype was built to produce a 300 mm stroke and facilitate the positioning of payloads of 0-10 kg with peak speeds of up to 1 ms^{-1}. The transmission system chosen was based on a lead screw which used a single re-circulating ball nut with a 20 mm lead on a 20 mm diameter shaft. A sketch of the completed module is shown in Fig 1.

The motor chosen for this system was an Electro-Craft brushed dc servo motor which had a continuous torque capability of .6 Nm (Peak 3 Nm). High quality brushed dc servo motors still offer the best performance/cost choice in a wide spectrum of application areas. However, for high performance applications where cost is not so critical then brushless dc motors offer a better choice (5).

3 DEVELOPMENT OF SINGLE AXIS CONTROLLER (SAC) - HARDWARE

The basis of the SAC in prototype form is a Texas 9900 single board computer. This was chosen mainly as appropriate development aids and considerable expertise for this processor and its family elements already existed in the Department. Furthermore the prototype "realtime" software can be transferred to 9995 based hardware with relative ease: this route being followed previously at LUT in producing pneumatic motion controllers in commercial form.

Control identification shows that both position and velocity are desirable parameters to measure (state variables). Position information has been obtained using an optical incremental encoder fitted to the drive. The resulting pulse-train is converted into absolute positional information by the use of a Texas LS2000 direction discriminator/counter chip. Velocity information traditionally is obtained by using a tachogenerator (tacho) which generates an analogue voltage which can be read by the microprocessor controller via an analogue to digital converter. However, tachos can add significant mass to the motor and since it is desirable to link a number of actuators together in robotic applications then it would be advantageous to find an alternative method of measuring velocity so that a tacho is no longer required: clearly if a tacho is not required potential cost savings can also be realised. Indirect velocity measurement has been achieved by sampling the change in position and/or determining the time period of pulses from the encoder (6, 7) thereby making the tacho redundant. The implementation of the pulse-width technique has been made simpler by the use of a LS2000 chip in pulse-width mode, combined with appropriate software. Using this method, velocity has been determined in the 3 to 3000 rpm range with an error which is never in excess of 2% of the true speed. Fig 2 shows an interface circuit suitable for use with the Texas 16 bit micro-processor which was designed to achieve indirect velocity measurement.

A schematic diagram of the complete system hardware is shown in Fig 3.

4 SOFTWARE

Of major significance in producing a flexible modular robotic system is the ability to define with ease the type of move and the control method by which the motion is achieved. The amount of human intervention (ie tuning of control parameters) and understanding of the workings of the motion controller must be kept to a minimum. The method of approach using command types, performance objectives and style have already been outlined in the introduction. Point to point motion and point to point with trapezoidal velocity profile have been achieved using a robust control approach (based on the theories of classical control) with good results. Self-tuning features which allow the controller to improve upon the performance of a robust controller are being investigated.

Essentially a robust controller employs a "de-tuned" control approach which can cope with plant parameter variations without causing unsatisfactory deterioration in performance. Load inertia represents probably the biggest system parameter variation in robotic work. Controller adaptation for this variable gives the ability to increasingly "tune" the controller in order to improve performance criteria. The process used to measure unknown plant parameters is called parameter identification. The methods by which plant parameter identification can be achieved are varied and is a subject of much academic interest (8). The method by which adaptation can be achieved is equally varied and there are many ideas, but few practical examples (9). However, one approach which has been widely accepted and used in servo control problems is the model reference adaptive control. The basic structure is illustrated in Fig 4. Essentially a model of the desired response is developed and this output (Ym) is compared to the process output (y). The error is then fed to an adaptation algorithm which in turn adjusts the coefficients of the feedback loop such that the plant follows more nearly the required performance.

This area of adaptive control (including self-tuning systems) is the area of present effort and it is anticipated that a design methodology will be identified and adaptive control systems impelemented and tested in the near future.

REFERENCES

(1) WESTON, R H and MORGAN G. A new family of robot modules and their industrial application. UK Robotics Res Conf, IEE, London, December 1984.

(2) McNAUGHTON, L S. Faster servodrives through pulse-width modulation. Machine Design, 1976, April 22, 57-61.

(3) LAIN, R N and HUGGINS, J A. State-of-the art in DC Motor Design. Proceedings of Motorcon, 1981, 392-405.

(4) NORCO ACTUATOR STANDARDS CATALOGUE, Norco Ltd, Industrial Estate, Waterford, Ireland.

(5) WYMAN, K. Choosing a servodrive - ac or dc? Electric Drives and Controls, 1984, October, 25-26.

(6) TAL, J. Velocity decoding in Digital Control Systems. Proceedings, 9th Annual Symposium on Incremental Motion Control Systems and Devices, 1980, 195-203.

(7) MOORE, P R, WESTON, R H, THATCHER, T W and HARRISON, R. Control of pneumatic servo drives using digital compensation. IASTED Int Symp on Telecoms and Control, Greece, August 1984.

(8) ASTROM, K J and EYKHOFF, P. System Identification - A Survey. Automatica, 1971, Vol 7, 123-162.

(9) JACOBS, O L R. Chapter 1, Self-tuning and adaptive control: theory and applications, Ed. C J HARRAS and S A BILLINGS, IEE Control Engineering Series 15, p 1-35.

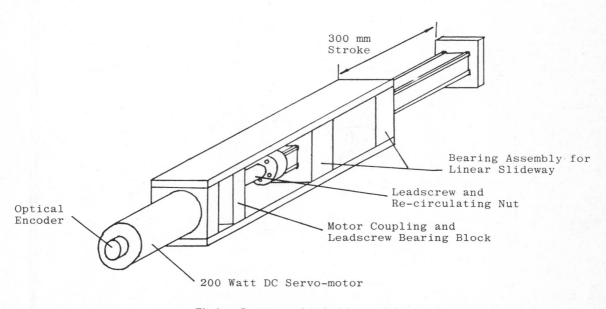

Fig 1   Prototype electric drive module

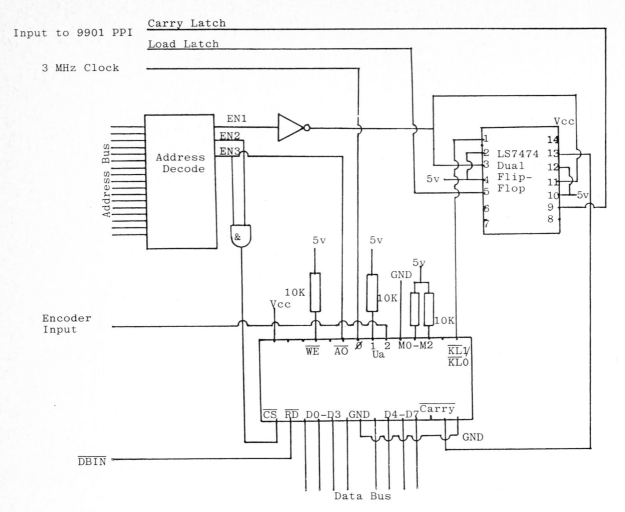

Fig 2   Circuit diagram of velocity decoder

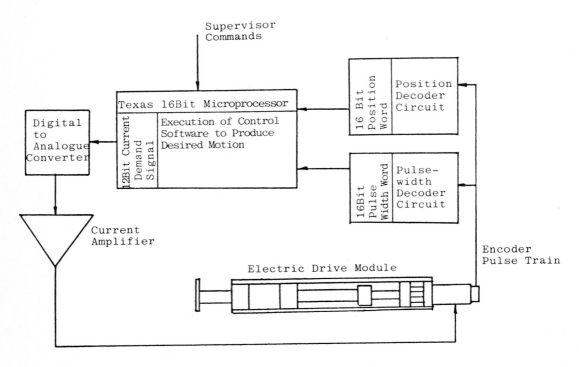

Fig 3   Schematic diagram of electric drive module hardware

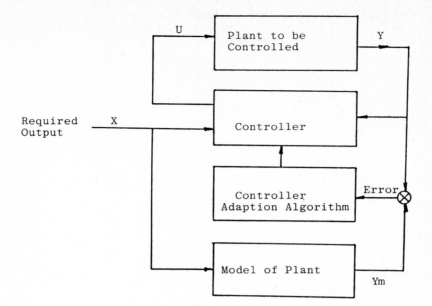

Fig 4   Model-reference adaptive control scheme

C371/86

Evaluation of a new approach to design and process planning

B S ACAR, MSc, PhD, K CASE, PhD, BSc, FErgs, MBCS, J BENNATON, PhD and N HART, BSc
Department of Engineering Production, Loughborough University of Technology, Loughborough, Leicester

SYNOPSIS

The analogy between the set theory operations in computer solid modelling systems and actual manufacturing operations provides an opportunity for the simulation of manufacturing processes at the design
stage. However, the mathematical terminology and procedures typically found in solid modellers make
them notoriously difficult to use in this context. Furthermore, with the help of a purpose-developed
system, much useful design information can be stored, to be used at later manufacturing stages.

This paper first describes the development of a CAD system, which provides an engineering interface
to a solid modeller and enables the creation of an outline process plan as the design progresses. An
evaluation of the capabilities of the system and the feasibility of the 'outline process plan'
approach to the planning problem is given. Future developments are proposed and examples from
industry are given.

1 INTRODUCTION

The creation of 2D detailed engineering drawings
at the design stage to transmit information to
the manufacturing areas has been a traditional
means of communication in industry. Recently,
2D CAD systems have been used instead of conventional means to generate the drawings, but their
successful interpretation still depends heavily
on the skill and judgement of the persons
involved between the design and manufacturing
stages.

Efforts are being made to automate the
process planner's role, and expert systems which
try to emulate the logic of a good process planner have recently been developed to overcome the
problems found with 'variant' and 'generative'
process planning systems (1,2). These systems
are generally based on feature recognition of
the designed engineering part using, for metal
cutting operations, the concept of 'recursive'
process planning, defined as the inverse
operation to machining. Using this technique an
operation sequence can be derived, working the
part from the finished state to the
configuration of the original material blank.

Such a procedure is clearly a reversal of
the design process, reconstructing from a
finished drawing the features which a designer
has encoded. This apparent repetition could be
avoided if the design and the process planning
stages are not thought of as two distinct ends
of the bridge from design to manufacture, but
considered as two essential components to
contribute simultaneously to generate the whole
process.

2 THE DESIGN AND OUTLINE PROCESS PLAN

Although it is expected that a good designer
will provide a product design which is both
functional and feasible to manufacture, it is
not typical for a designer to have had extensive
training and experience in manufacturing processes. However, this problem would be considerably eased if the required manufacturing knowledge is provided by a purpose-developed system.

The advent of computer solid modelling systems provides an opportunity for the simulation
of manufacturing processes as an aid to design.
Some solid modelling CAD systems use a method of
geometry specification which has a close resem-
blance to certain manufacturing operations. It
is beneficial not to waste the valuable information dealt with while the component is being
designed (3,4). However, at present it appears
that no system provides a mechanism for the
retention of the considerable amount of geometric information produced about the simulated
manufacturing operations. Furthermore, the
handling of the 3D information in these systems
is typically carried out in mathematical rather
than engineering terms. Typical command
languages require a certain level of knowledge
of Axiomatic Set Theory, Modern Logic and
Euclidian Space Norms.

The system described below has been
developed for use as a tool for building 3D
models of engineering parts, on a graphics
terminal, using familiar engineering language,
and to produce their associated outline process
plans.

3 DESCRIPTION OF THE CAD TOOL

The system is based on the concept that manufacturing operations are analogous to the basic set theory operations. The development of the prototype system has concentrated on metal cutting operations, in particular on turning, milling and drilling operations, since these are widely used in engineering and allied industries.

Metal removing operations are defined as the difference between the set of points of the material and the set of points created by the tool movement. The latter set is obtained by applying set theory operations such as union, intersection and difference, to the sets of points of some primitives - blocks, cylinders, toruses and cones; whereas the former set is a set of points in a block or any other preformed shape. For example, the action of a chamfering tool sweeping along an edge of a rectangular block is simulated by the difference between a parallelepiped and the union of suitably positioned, truncated and suitably rotated parallelepipeds and two truncated inverted hemi-cones of appropriate dimensions.

3.1 Input

Stock material is taken in the form of a rectangular plate or block, or alternatively a preformed shape may be retrieved as the stock material. In the latter case, the definition of the shape is created using the command language of the solid modeller.

The list given in Table 1 contains available machining operations (1-12) and some other facilities (13-19) of the system.

Table 1 List of machining operations and general facilities

1	DRILL HOLE	11	KEYWAY
2	DRILL&BORE	12	FILLET
3	DRILL&REAM	13	ROTATE
4	DRILL&TAP	14	REVOKE
5	COUNTERSINK	15	EXAMINE OL.PROCESS PLAN
6	BORE/COUNTERBORE	16	DRAWING ON
7	COPY(HOLE)	17	DRAWING OFF
8	STEP	18	VIEW MODEL
9	CHAMFER	19	EXIT
10	SLOT		

Selection from screen menus and the use of graphics facilities such as cursor input to indicate reference points, milling directions, etc, are included as part of the user interface for convenience. However, the information needed to describe the mathematical model to represent the manufacturing operation is in the form of a language command with the following structure:

APPLY OPERATION $(D_1,..,D_k)$ AT (X,Y,Z)

where

APPLY OPERATION is the engineering term for the required operation, eg MILL STEP, DRILL HOLE, BORE HOLE.

$D_i, i=1,..,k, k \in \mathbb{N}$ is the required parameters of the feature, eg length, width and depth of a step; diameter and depth of a hole.

(X,Y,Z) represents the co-ordinates of the reference point in $\mathbb{R}^3$.

The programs are written to translate the information into a suitable format in a solid modeller, which is provided at present by the subroutine version of BOXER (5).

3.2 Procedure

At the end of each operation the resulting worked piece is taken as new material ready to be worked. The system is capable of implementing up to 99 operations on the same piece, provided sufficient memory is available.

The model should be ROTATED to be able to apply machining operations to the sides or bottom face of the part, since the system is at present intended for vertical axis machines. The REVOKE facility can be used to cancel unwanted operations. It is possible to EXAMINE the OUTLINE PROCESS PLAN which is being generated during a design session and contains much of the manufacturing information (see 3.3 below). The graphics facility is available to display the shape of the part after each operation unless the DRAWING facility is turned OFF, although it is turned ON automatically when it is needed. Other VIEWing facilities are available through the normal interface to the solid modeller.

The system is user friendly, therefore some restrictions are imbedded to prevent the user designing physically impossible parts. The user is also forced to follow some simple manufacturing rules, such as creating the holes before countersinking or counterboring them. The development of an expert system approach to the monitoring and guiding of the user's actions in relation to established design and manufacture rules is the subject of future research.

3.3 Output

The output consists of two main parts:

(i) A 3D graphical view of the generated component.

(ii) An outline process plan file for the component in the form of a Table of manufacturing information. This includes the operation number and type, the dimensions of the generated feature, the specifications of a default tool type, the co-ordinates of the initial tool position and the direction of the movement of the tool along an axis (tool path) to carry out the operation.

A combination of these two outputs, graphical views of the part for various stages of manufacture, together with the corresponding text extracted from the outline process plan file, are also available.

4 EVALUATION OF THE CAPABILITIES OF THE TOOL

To assess the capabilities of the software to simulate the production of typical engineering parts, information in the form of drawings and process plans was obtained from two collaborating companies. The parts were selected as those produced by metal cutting operations and a sub-sample of these was used for the evaluation.

The combined outputs presented in Figure 1 and Figure 2 are produced from sample parts and show graphical views with associated text, which represents a general description of each manufacturing operation.

4.1 Examples

The samples obtained from the companies were classified into groups of simple and complex parts according to the manufacturing operations required to create them. To assess the capabilities of the system, the first example is chosen from the complex group (since the success of the system on this group would imply success on the simple group) and the second example is chosen from the simple group (to exhibit the usefulness of some of the facilities).

Figure 1 shows a possible sequence to manufacture Example 1, an Arm Folder taken from Baker Perkins, Peterborough. As seen in the Figure, rotations are essential for some operations at the sides or bottom face of the component. The exhibited sequence minimises the number of rotations, ie every manufacturing operation is applied, once the material is in the required position. Obviously the outline process plan can be optimised by changing the order of the operations, sometimes with additional rotations.

The stages of designing Example 2, a Gearbox Sideplate taken from Herbert Morris, Loughborough, are shown in Figure 2. Unlike Example 1, the stock material is taken as a flame cut contour which is previously defined and stored in a file. This example proves the usefulness of the COPY(HOLE) facility. The pictures are given in 2D, since all the operations are performed on the top of the part, ie the rotation facility is never used.

It appears that the quality of the design depends on the information that can be given to a user related to geometry, machines, standard tooling and capabilities of manufacturing processes. This is the subject of future developments.

4.2 Experiments with novice users

Some experiments have been designed to investigate the acceptability of the approach and the system. Groups of final year Production Engineering students at Loughborough University of Technology were used to make some preliminary evaluation studies, although the non-randomness of the sample of users is acknowledged. Analysis of the results of the novice user experiments shows that:

(i) The system is very easy to use.

(ii) Outline process plans are much clearer than traditional process plans.

(iii) It is possible to obtain a 'good process plan' if the users employ their knowledge of manufacturing processes.

4.3 Reactions of the potential users

The reactions of the designers and engineers in industry to this novel approach have yet to be fully investigated. It is not very surprising to receive a somewhat 'sceptical and reluctant' reaction, which is summarised below in the responses from engineers from the collaborating companies:

(i) Design, Process Planning and Manufacture are three clearly defined areas or departments. The use of our tool would imply a revision of certain procedures within these initial manufacturing functions.

(ii) The means of constructing a model as a series of manufacturing operations and not in terms of 'functional' design features could lead to difficulties in acceptance of the system by designers.

(iii) The cost of 3D modelling CAD systems can still be significant.

Experiments, hands-on sessions, are currently being designed for the designers and engineers from industry to complete the evaluation of the system.

5 DISCUSSION

Simulating manufacturing processes at the design stage may be an answer to the 'design for manufacture' concept in a truly integrated CAD and CAM system. The approach implemented here provides an alternative way to obtain a process plan and avoids the complex logical rules required to develop a generative process planning system.

The system enables the designers to use a solid modeller and 3D representation facilities without worrying about the mathematical complexities of the command languages.

The outline process plan, by definition, is not necessarily the optimum sequence of operations to manufacture the component. However, it is at least an available possible sequence of operations which can well be optimised by re-ordering the operations at the next stage, according to the priorities of the manufacturer.

Preliminary evaluations of the prototype CAD tool based on a solid modeller have been made by using typical parts from two engineering companies. A number of advantages have been identified for the use of the tool in the design and process planning areas. Various immediate reactions to its implementations have also been acknowledged. Further future developments have been proposed in the light of the results of the experiments.

REFERENCES

(1) DAVIES B J, DARBYSHIRE I, WRIGHT A and PARK M W. The integration of process planning with CAD/CAM, including the use of expert systems. Proc International Conf on Computer Aided Production Engineering, University of Edinburgh, 1986.

(2) MATSUSHIMA K, OKADA N and SATA T. The integration of CAD and CAM by the application of artificial intelligence techniques. Annals of CIRP, 1982.

(3) BENNATON J, CASE K, HART N and ACAR S. A system to aid design by planning manufacturing operations. Proc 1st National Conf on Production Research, University of Nottingham, 1985.

(4) HART N, CASE K, BENNATON J and ACAR S. A CAD language to aid manufacture. Proc International Conf on Computer Aided Production Engineering, University of Edinburgh, 1986.

(5) BOXER Solid Modelling System, Level 3.2, Subroutine Driven Version. PAFEC Ltd, Nottingham, February 1986.

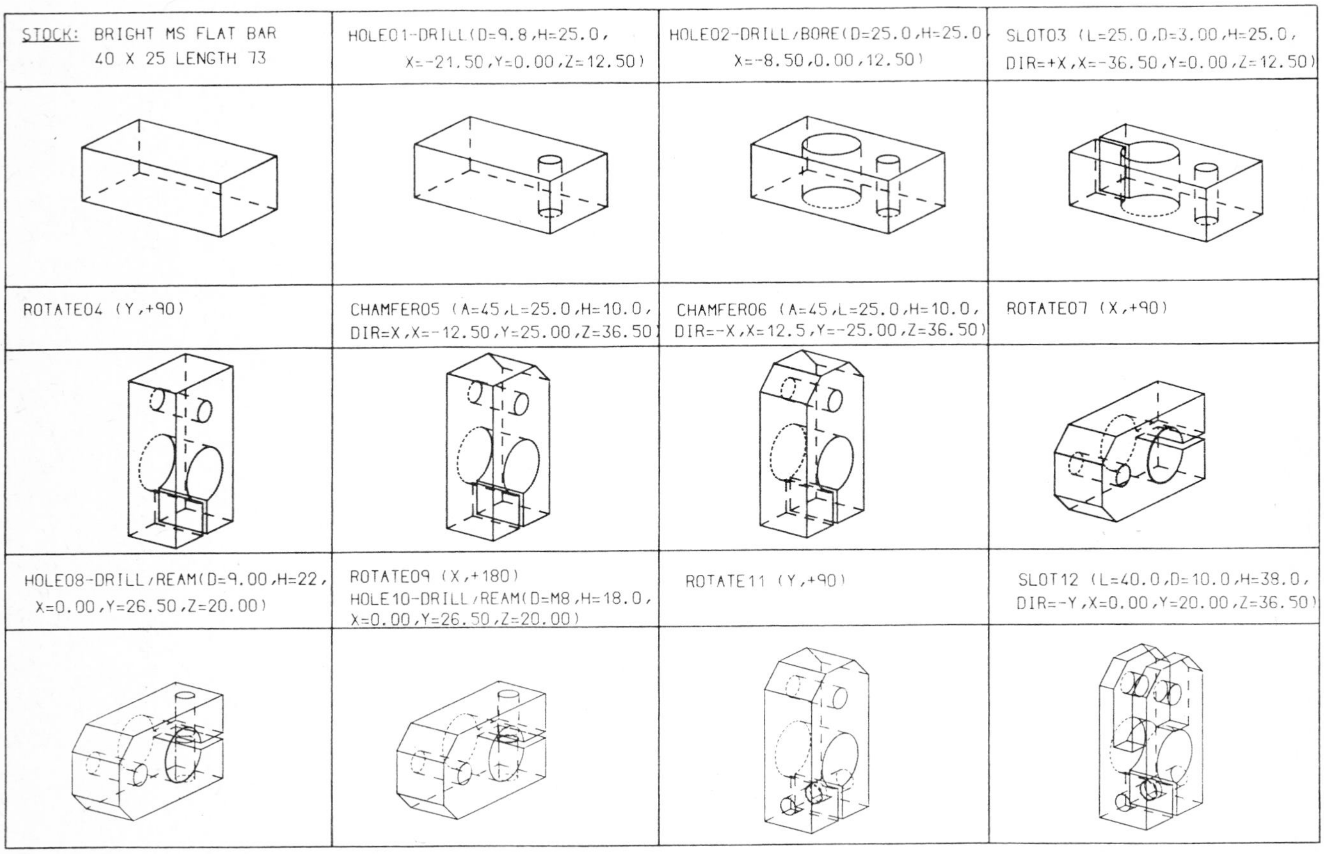

Fig 1 Outline process plan for a part created using various milling operations

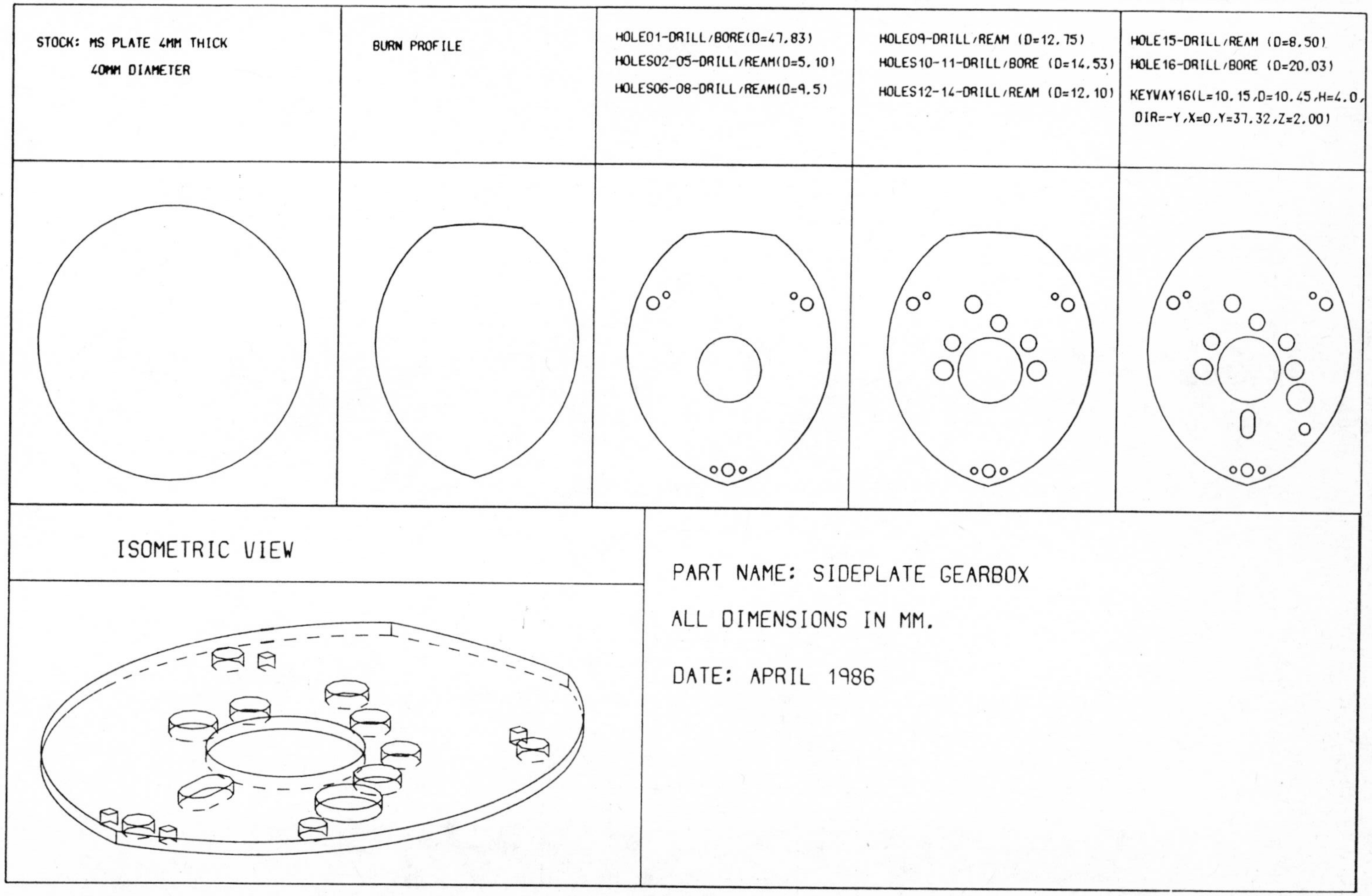

STOCK: MS PLATE 4MM THICK
       40MM DIAMETER

BURN PROFILE

HOLE01-DRILL/BORE(D=47.83)
HOLES02-05-DRILL/REAM(D=5.10)
HOLES06-08-DRILL/REAM(D=9.5)

HOLE09-DRILL/REAM (D=12.75)
HOLES10-11-DRILL/BORE (D=14.53)
HOLES12-14-DRILL/REAM (D=12.10)

HOLE15-DRILL/REAM (D=8.50)
HOLE16-DRILL/BORE (D=20.03)
KEYWAY16(L=10.15,D=10.45,H=4.0,
D1R=-Y,X=0,Y=37.32,Z=2.00)

PART NAME: SIDEPLATE GEARBOX

ALL DIMENSIONS IN MM.

DATE: APRIL 1986

ISOMETRIC VIEW

Fig 2   Outline process plan for a part machined from a preformed material

C377/86

The development of a pneumatically powered walking robot base

A A COLLIE, BSc, CEng, MIEE, J BILLINGSLEY, MA, PhD, CEng, FIEE, FIOA and L HATLEY, BSc
Department of Electrical and Electronic Engineering, Portsmouth Polytechnic

SYNOPSIS   The development of a six-legged walking robot base unit powered by compressed air is described. It uses adaptive control to adjust its gait according to the terrain. It is designed to carry special purpose tools for utilization on site.

1   INTRODUCTION

The concept of a multilegged walking robot does not at first seem to be a serious one, but such devices have real applications. Groups in Russia, the USA and Japan are actively working on walking machines. Production of Odectic's robot is well established in the USA, Hitachi have a sixteen legged crawling robot currently on trial inspecting large spherical pressure vessels.

Most industrial robots are bolted down devices with a moving limb mounted on a rigid pedestal. They are designed to work in the typical factory environment where the products are moved from one work station to the next. Since most robot designers work in a similar environment, their robots tend to perpetuate the conventional factory concept of large concrete floors covered by low buildings. The larger the component being made, the larger the building and the need for transport mechanisms to carry the product around the factory. Of course productivity can be increased by conventional robot assembly techniques, but on a national scale there is a limit to the benefits which are to be obtained. Most people do not work in factories.

It is the authors' belief that there is a need for a low cost mobile work platform capable of sharing man's environment and which can take the process to the product. Wheeled robots have started to appear for transporting products around factories, running on carefully prepared surfaces. Most working environments do not have a smooth, firm, level floor.

The mobile robot walking base described here is able to carry a variety of special tools and equipment intended to increase the productivity of the majority of the population who do not work in a special environment. Agriculture, at one time employed 95% of the British population: machinery and modern methods have produced a surplus using the efforts of less than 5%. Improvements of a similar nature might be made in industry at large.

Shipbuilding, Mining, Construction and Building could all benefit from a machine which could manoeuvre on broken ground, negotiate loose footing and climb stairs or over obstacles. The need for small, semi-autonomous mobile radiation-proof robots for repair or decontamination work in the nuclear industry is self evident.

Tracked or wheeled vehicles could carry out much of this work but a walking machine has clear advantages which justify its additional compli-cation. It can feel its way, only transferring its weight when it has tested its footing. This gives it the ability to surmount obstacles out of all proportion to its size. Tracked and wheeled vehicles, on the other hand, can be seriously impeded by gullies of about half their size.

The previous approaches to walking devices have concentrated on linkage and telescopic mechanisms, driven by electric motors or hydrau-lic drives. These have produced large and power-ful machines which however, are geometric in their nature. That is to say that at any point in its stride the machine forms a rigid structure adopt-ing a stable position as it rests on the support-ing surface. The mechanism described in this paper approaches the task in a more general way, employing compliance and adaptation to achieve smooth fluid motion over uneven terrain by means of jointed limbs powered with compressed air.

2   GAIT AND COMPLIANCE

A conventional wheeled vehicle is supported on a mechanical suspension, which deflects to accommodate defects in the surface, but which will nevertheless cause the body to perform small vertical and angular excursions. The mean position is dictated by the mechanical structure of its suspension. In exceptional cases of rough ground, one wheel or a pair of diagonally opposite wheels may leave the ground, leaving in the latter case one degree of freedom to be determined by the dynamics of the vehicle. Suspension design then entails a compromise involving time-constants, damping factors, excursions of mech-anical members and limiting values of ground clearance on a variety of surfaces.

The body of the walking robot will be similarly subjected to the reaction forces of the legs. Each leg in contact with the ground will afford lift to the body, dictated not just by deflection geometry and springing as in the case of a wheel but in accordance with an actively controlled force. With gravity, these are the only forces acting on the body, apart from wind loads and such disturbances. In theory, the body can thus be "flown" above the ground in a deterministic manner, subjected only to those accelerations necessary to guide its path.

The legs must hold the body clear of potential obstructions, but a gait which seeks excessive height will decrease the possible positional range of the ground contact points - in the extreme case when the height is equal to the leg length, each foot touching the ground must be immediately beneath its hip. In practice, compromise must be accepted between speed and evenness of travel. Within the limitations of the performance and response of the legs, different gaits become appropriate at different speeds.

The simplest gait is a "prowl" in which the centre of gravity of the body is not permitted to move out of the area enclosed by the ground contact points. The velocity is slow enough that the machine can stop at any point in a stable state. Next comes a "scurry", in which the centre of gravity is still constrained as before, but where the speed may demand one or more paces to come to a halt. In these gaits, the body can move smoothly in a horizontal line without angular disturbances. The resultant of the leg forces can satisfactorily be constrained to be a vertical force through the centre of gravity. There is an overriding constraint on the proportion of the time which each leg must spend in contact with the ground, and a limit on the possible length of stride.

The speed may be pushed higher by allowing temporary instability, where the centre of gravity passes outside the leg base. The body motion may then contain periodic elements of linear or angular acceleration - or both. Constant velocity of contact feet relative to the body will give smooth linear motion, with a lurching rotation as toppling instability occurs and is corrected - this is best likened to a "canter". On the other hand, a "lope" can allow the vertical foot reaction to combine with an accelerating or decelerating force to give a resultant through the centre of gravity. The motion will then be free from rotational disturbance, but will contain horizontal accelerations. The motion can instead contain vertical accelerations, a "trot", as a consequence of varying vertical foot resultants.

In a "gallop" all feet may spend a substantial part of the cycle off the ground, leaving the body in free fall. Although uncomfortable and involving high leg stresses, this may still be a desirable option to escape danger. In any rapid gait, the inertial dynamics of the legs may themselves play a significant part in balancing or disturbing the vehicle.

3   STRUCTURE

Six legs are mounted on a roughly coffin shaped body about a metre long and standing just over half a metre high. Six legs are used, since with only four there are severe constraints on leg motion geometry for slower gaits in which those in contact with the ground must provide a structural base. Five legs seemed inappropriate!

Figure 1 shows the leg mechanism. In the thigh two single acting cylinders connect via a steel tape to a drum attached to the lower leg. The two cylinders act in opposition so that a differential pressure between them will apply a force to the lower leg. If air is let into both cylinders simultaneously, however, the effect is to damp the movement and stiffen the limb. The upper leg is operated by a double acting cylinder, connecting rod and crank. Less angular movement is needed at the hip and this scheme saves space. The crank can be offset relative to the thigh so that maximum force can be applied at a particular part of the range to suit an application.

The hip joint has two degrees of rotational freedom, allowing both a long fore and aft stride (extension and flexion) and a short sideways movement (abduction). Abduction is a complication but is necessary to increase the manoeuverability required of a robot designed to position tools or equipment in confined spaces. To limit the complexity in this model the sideways stride has a fixed length of 100 mm.

A walking machine can be steered in a number of different ways, only a few of which have been considered. The mechanism is adaptable, and alternatives can be investigated later. For small changes in direction the stride on each side of the machine can be lengthened or shortened. To assist this effect the axes of the hip joints are inclined so that the projection of the locus of the foot onto the ground is a curve. The mid leg on each side has its hip inclined in the opposite sense to the extreme legs and to a slightly greater amount so that a common turning centre is produced. A second advantage of this geometry is that it enables the mid legs to pass outside the end legs when an overlap occurs. For more rapid turns the stance on the inside of the curve can be shortened. In the limit, of course, the robot can pivot around one stationary leg.

4   POWER SOURCE

The choice of compressed air seems at first sight to be an unsatisfactory method of powering the limbs, but study of walking motion shows that the important characteristics are that the system is resilient when moving but capable of locking rigidly when stationary. These are inherent properties of pneumatic actuation whereas with electro-mechanical actuation a damped spring suspension becomes necessary, with its consequent increase in complexity.

Hydraulic actuation is a good alternative choice: resilience can be included by air accumulators but the prime mover must then be carried with the vehicle. Since one of the objectives is to achieve a low manufacturing cost, hydraulic actuation seemed more appropriate for a larger machine. Further, in the intended application an air supply is normally available, and the cost of a prime mover is avoided. A trailing umbilical cord need not be a disadvantage. It may in any case be essential to carry remote control and closed circuit television signals. In those applications where no umbilical cord can be used there have been some recent developments in very high pressure air storage cylinders which (calculations show) can give the robot either a short gay life of 100 m in 10 sec or a slow longer one. Of course, a combination of an air reservoir and a portable compressor for topping up is the ideal solution for a free ranging machine.

Using compressed air in a servo mechanism presents special control problems but a walking movement is a dynamic one in which great positional accuracy is not required. In fact over rough ground the robot has only an approximate "notion" of where the ground is. It is the forces exerted to provide forward drive and support the weight of the complete machine and load which are important. Figure 2 shows the basic control system for one leg. Each leg has its own single board computer which pulses pairs of very fast acting diaphragm valves connected to each pair of opposing cylinders. The pairs of valves are arranged so that one is an inlet and the other an outlet, thus by giving different pulse patterns to the valves the pressure in each cylinder can be accurately controlled. Silicon sensors monitor the pressures resulting from the pneumatic forces and the ground reaction while the actual joint angles are measured by potentiometers. The burden of the computer is reduced by an analog inner control loop which seeks to maintain the pressure required, leaving the computer to solve the joint angle relationship and calculate the pressure demand.

Figure 3 shows a graph of a saw tooth displacement profile produced by controlled air pulses. The pulse wave form is here continually adapted to hold the displacement profile within limits.

5 LEG CONTROL

Commands to the leg computers are given from the central controller which coordinates the movement pattern in terms of STANCE, PACE, STRIDE and GAIT. The central controller also takes account of the terrain and computes the "Dead Reckoning" position.

Figure 4 shows a limb in mid stride from which the following relationships are derived.

$$X = L\mathrm{Sin}\theta_1 + L\mathrm{Sin}(\theta_1 + \theta_2)$$

$$Y = L\mathrm{Cos}\theta_1 + L\mathrm{Cos}(\theta_1 + \theta_2)$$

For robot kinematics, Paul's notation has become widely accepted, representing $\mathrm{Sin}(\theta_1 + \theta_2)$ more concisely as S_{12} etc.

Now

$$X = Ls_1 + Ls_{12}$$

$$Y = Lc_1 + Lc_{12}$$

where L is the length of the upper and lower leg, designed to be equal.

If W_r is the Walking Resistance and M the mass of the robot and its load per limb, then the torque at the joints is given for the hip as:-

$$T_{hip} = M g X + W_r Y$$

$$T_{hip} = M g L (s_1 + s_{12}) + W_r L (c_1 + c_{12})$$

and the knee as:-

$$T_{knee} = M g (X - Ls_1) + W_r (Y - Lc_1)$$

or

$$T_{knee} = M g L (s_{12}) + W_r L (c_{12})$$

from which the pressures can be derived.

$$P_{hip} = M g K_h (s_1 + s_{12})/c_1 + W_r K_h (1 + c_{12}/c_1)$$

and

$$P_{knee} = M g K_k (s_{12}) + W_r K_k (c_{12})$$

where K_h and K_k are constants dependent on dimensions.

The values of the hip and knee angles are measured and hence the above pressures can be worked out directly from a look-up table of the sines and cosines of angles. For the accuracies required it is sufficient to work to the nearest degree.

Each pace is made up of two distinct parts. The free forward extension stroke, when the leg is not touching the ground and the power flexion stroke when the leg is weight bearing and imparting horizontal acceleration.

One of the problems associated with moving, jointed, limbs is the inertia forces transferred as each part is independently accelerated and decelerated. Traditional control techniques have implemented non linear transfer functions in which damping and loop gain are made to change with the relationship between each part of the limb. The method described here avoids this complication by using an adaptive control algorithm which adjusts its responses over a number of strides to optimise the performance.

Extension Mode 1

During Extension the limb is made to follow one of a number of movement profiles which are held in memory. These define the approximate knee angles for corresponding hip angles appropriate to the stance and stride. They define "target" points which would just keep the foot clear of the ground. In this mode the robot is able to feel for any obstructions and learn to avoid them.

At the start of a stride a strobe time at which each target should be achieved is calculated. Next the appropriate inlet valves are given a pulse of air whose duration is proportional to the angle between the start position and the first target. At the next strobe a further air pulse is given whose duration is proportional to the difference between the current position and the third target, meanwhile noting the error between the current position and the second target. This error is held in temporary memory and a fraction of it is added to the first air pulse of the following stride. This procedure repeats for each target until the end of the extension is near. At this point inertia forces are more significant and the procedure is slightly modified by closing the outlet valve of the opposing cylinder early to obtain some energy recovery. The point at which this and the start of Flexion takes place is determined over a number of strides using a similar error correction procedure.

Clearly, for a normal stride on level ground, there is a consequent pressure profile corresponding to the position profile. This is more or less the same for each stride, however, if the movement is impeded the pressure in the cylinders will increase. (The control algorithm enhances the effect). By comparing the current pressure profile, position by position with the standard, obstacles can be detected. The action the computer takes is to select a different set of values from the table of movement profiles corresponding to a lower stance. Since the hip remains at a constant height above ground the foot lifts and the pressure is released as the movement continues. During subsequent strides a higher stance is tested and the new profile is either reduced or confirmed. The objective is to find the lowest profile consistent with minimum air consumption.

When an obstruction is detected its coordinates are mapped in cartesian form by the relationships above and the information is passed to the central computer to enable following limbs to anticipate the obstruction and so improve the gait. Over the next year, work concerning the central control computer will be centred on recognising terrain patterns such as stairs, inclines or loose footing. This will require additional instruments for accurate control.

Flexion Mode 2

During Flexion, the limbs bear weight and force balance control is used, part digital and part analog. Figure 5 shows the block diagram of the analog inner loop. The air valves respond to the difference in pressure between the demand signal from the computer and the pressure feedback from the sensors. A phase advance network provides stabilization. A gate on the output directly controlled by the computer can override the analog loop when required.

Flexion starts at the end of Extension when the foot may still be above the ground. The procedure starts by setting up a flexion strobe timer and is initiated by a phasing signal from central control to ensure limb coordination. The leg computer then evaluates the pressure demand for the two flexion cylinders needed to support the weight and overcome the walking resistance at the current hip and knee angles. When the actual pressure matches the demand then the foot has touched the ground. At this time the new values of X and Y are evaluated and sent to the central control for subsequent correction, if required, at the next stride.

As each strobe pulse occurs the current vertical and horizontal displacements are deduced. Comparison is made with the stance and pace set by the command, and a correction is made to the weight or walking resistance term of the equation if needed. A permanent adjustment is only made to the current term if the current error is greater than a given threshold and the previous error is less than the tolerance.

6 INSTRUMENTATION AND CONTROL

Only very limited provision can be made to displace the legs to the side. Any rolling motion of the body must be corrected by differential leg reactions; a relatively modest disturbance can become irrecoverable. Rolling must thus be corrected swiftly, implying a need to detect first or second derivatives of the disturbance. Rate gyroscopes or equivalent instruments will be needed for stabilising any rapid gait. Since long-term integration is unnecessary, low-cost devices can be used despite their liability to drift. By monitoring the leg forces and performing appropriate computations, roll acceleration can be estimated, avoiding the need for additional instrumentation. Attitude can be readily deduced from a set of linear accelerometers, and the computer is again used to monitor and compensate for the limitations of the instruments. Most important is the measurement of leg attitudes and the forces acting on them. The pressures within the pneumatic cylinders are already measured for closing high-speed control loops around them, and these afford variables for force determination.

For each leg, three angles are measured – the knee and thigh angles, and the abduction angle of the hip joint. From these the position of the foot can be calculated relative to the body as r_x, r_y, and r_z, where direction x is taken as forward, y to the left and z upwards, relative to body axes aligned with the principal axes of inertia. Three sets of pressures are also measured, giving three torques from which can be deduced the three components of the reaction force at the foot. If we denote these by $F_{i,x}$, $F_{i,y}$ and $F_{i,z}$ for leg i, then the following equations describe the body motion:

$$M\ddot{x} = M\underline{g} + \Sigma \underline{F}_i$$

$$I\ddot{\underline{\theta}} = \Sigma \underline{F}_i \times \underline{r}_i$$

where I is the inertia tensor.

For smooth, irrotational motion we require:

$$\Sigma \underline{F}_i + M\underline{g} = 0$$

and

$$\Sigma \underline{F}_i \times \underline{r}_i = 0$$

Although at the time of writing only one leg has been fully developed, work has been put in hand to construct the entire machine. By the time of presentation the device will have been commissioned and evidence can be shown of its performance.

General Notes and Definitions

Central Controller

responds to high level commands
e.g. forward, left, slower, etc.

instructs each leg with an intention command consisting of: stance, gait, pace and stride.

decides action on information returned from legs.

keeps track of dead-reckoning (distance and direction (vector), and velocity).

Leg Controller

responds to information learned from actual, as opposed to intended movement

controls action of leg
sends information about terrain to central controller

Definitions

Stride

The ratio length of stride to leg length.

Stance

The ratio of height of hip to length of leg.

Gait

Controls the relationship between knee and hip angles. This defines the type of movement, e.g. WALK, GROPE, GALLOP.

Pace

The number of paces per second.
(Note: speed = Pace * Stride * Leg Length).

Mode

The way in which the movement control algorithm operates, e.g. Mode 1 is Time/Displacement, Mode 2 is Force Balance.

REFERENCES

(1) THRING, M. Robots and Telechirs. Ellis Warwood Ltd., 1983, Chapters % & 6.

(2) RAIBERT, M.H. and SUTHERLAND, I.E. Machines that walk. Scientific American, Jan. 1983, 32-41.

(3) WESTON, R.H., MOORE, P.R., THATCHER, T.W., and MORGAN, G. Computer controlled pneumatic servo drives. Microprocessors and Fluid Power Engineering, I.Mech.E. Conf. Publication, 1984, 8, 97-106.

(4) TODD, D.J. A novel steering mechanism for legged robots. Robots and Automated Manufacture. Peter Peregrinus Ltd. for IEE Control Eng. (Series No. 28), 1-10.

(5) CHAPMAN, R.F. The insects, structure and function. The English University Press, 1969, 127-164.

(6) TAYLOR, R. Force development during sustained locomotion: A determinant of gait, speed and metabolic power. Journal of Experimental Biology, 115, 253-262.

(7) McMAHON, T.A. The role of compliance in mammalian running gaits. Journal of Experimental Biology, 115, 263-282.

(8) PAUL, R. Robot manipulators, MIT Press, 1982.

(9) FUJITA, A., TORGE, M., MORI, K., SONODA, S., WATAHIKI, S., OZAKI, N. Development of inspection robot for spherical gas storage tanks. Proceedings of the 16th International Dymposium on Industrial Robots, 1185-1194.

The authors would like to thank the Science and Engineering Research Council, the Royal Society, the Royal Armoured Research and Development Establishment, Tube Investments and the West Group for their support and encouragement. They would also like to thank Dr. Wakely and Professor Thring for their helpful discussions at the start of the project.

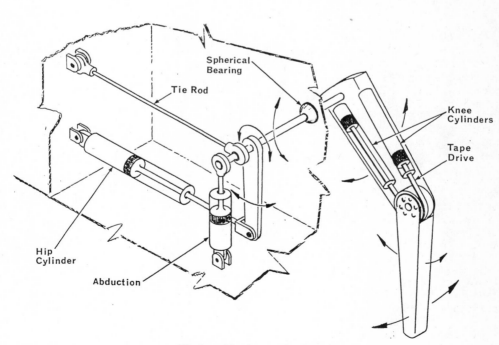

Fig 1   The leg mechanism

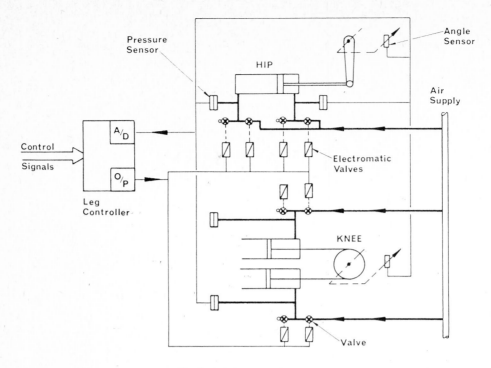

Fig 2 Leg control system

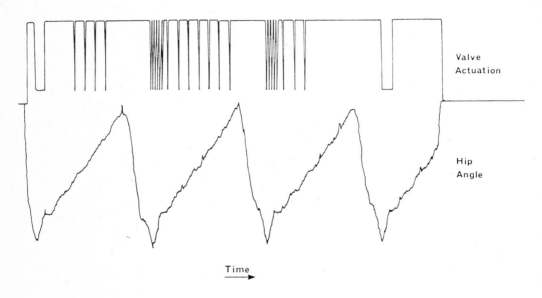

Fig 3 Adaptive control for saw-tooth displacement

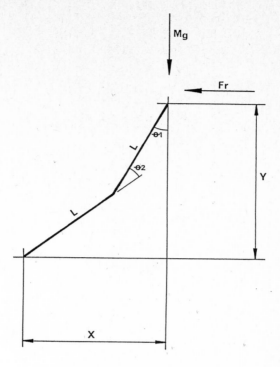

Fig 4   A limb in mid-stride

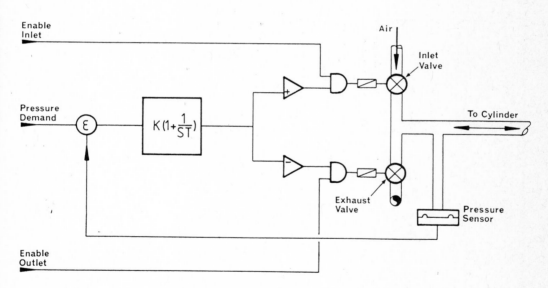

Fig 5   Inner loop control

C378/86

Implementation of adaptive positional control and active compliance for robots

M SAHIRAD, MEng, BSc(Eng), **M RISTIC**, PhD, BSc(Eng), MSc, DIC and
C B BESANT, PhD, BSc(Eng), DIC, CEng, FIMechE
Department of Mechanical Engineering, Imperial College of Science and Technology, London

SYNOPSIS Direct drive robots have no gearing between the actuators and the joints, so it is possible to construct an actively compliant robotic system which does not need any specialised force sensors. On the other hand, the absence of gearing makes the overall system much more non-linear and difficult to control during high-speed motion. This paper describes the strategies needed to control such a robot. It is shown how active compliance can be achieved by means of measuring the positional error in the joints and how dynamic compensation can be achieved via adaptive control techniques to give accurate high speed trajectory following.

1 INTRODUCTION

With the increasing acceptance of industrial robots by the industry the sophistication of the tasks that they are demanded to perform has increased too. In particular the emphasis has been shifted from the simple pick-and-place operation to automating assembly tasks, in which mating of closely fitting parts is required. Because the robots have poor ability to deal with task uncertainties, current installations typically rely on the extremely high positioning accuracy of the robot arm and on the accurate positioning of the workpiece itself. Thus the working environment has to be highly structured and invariable which means that robotic assembly is difficult and expensive to set up, while the manufacturing tolerances make every assembly operation unique.

In order to guarantee the positioning accuracy of the overall system, the conventional robotic arms are designed to be very stiff but often it is the inherent and *uncontrolled* compliance that allows the assembly operation to proceed at all. Gripper elasticity, servo gain and structural stiffness all contribute to the intrinsic compliance in a task. Paradoxically, they aid assembly by accommodating a certain degree of part misalignment.

In order to aid assembly further, specialised devices (Remote Centre Compliance) have been used to introduce an additional amount of compliance in the system, but setting up of such a system is highly operation specific.

Sensory feedback is an obvious answer to the problems of robotic assembly. While vision is considered to be of limited value in this context due to its limited resolution, properly used force feedback seems to be much more important. Robotic force sensors have been used with success, but their use implies an increase in the overall complexity and price of the system and a decrease in reliability. In theory force feedback could be provided by measuring the current supplied to the actuators, but with high gearing ratios of conventional arms (typically 200:1) and the associated friction this becomes very difficult to achieve in practice.

2 DIRECT DRIVE ARM

In an attempt to reconsider the whole design philosophy of assembly robots a lot of work has recently been done on the development of Direct Drive (DD) robot arms. A DD arm is a

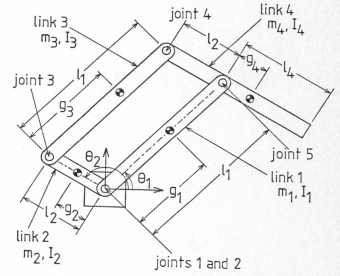

Fig 1　Two degree-of-freedom direct drive arm structure

mechanical arm in which the joints are directly coupled to the rotors of high-torque/low-speed electric d.c. servomotors. Since the arm does not contain any gears or transmission mechanisms between the motors and their loads, the drive systems have no backlash, small friction and high mechanical stiffness, all of which are desirable for fast, accurate and versatile robots. When compared to the conventionally designed robots, DD robots have the following advantages:

- High speed and acceleration
- Compliance can be easily controlled by varying controller gain
- Force exerted by the gripper can be easily related to the current in the motors

With these qualities in hand it becomes possible to construct a reliable and inexpensive high performance robotic system. In such a system the robotic arm would be designed to be light and fast but also simple in construction while the complications of sensory control will be taken on by the controller software. In this way the system would not have to rely on sophisticated sensors for its operation.

This paper deals with the control strategies required for such a system.

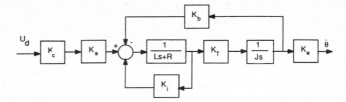

Fig 2   Block diagram of the single joint transfer function

A direct drive arm of the SCARA-type, with parallelogram linkage mechanism, has been constructed at Imperial College for experimental purposes. A second arm is in the final stages of its design. It is intended to manufacture the linkages of the new arm from carbon fibre based materials which would give rise to an even lighter arm and consequently faster dynamic response. The fibre based material would also dampen out any mechanical vibrations.

It is assumed that the robot will be required to perform two basic kinds of motion:

1. Compliant motion: end effector exerts the required forces on the workpiece during the motion.
2. Motion in free space: the requirement is to follow the given spatial trajectories accurately and at high speed.

Conversely, the control strategies to suit the different operations will be presented.

3.1   Invariant and decoupled arm inertia

In a conventional serial linkage arm, dynamic interactions exist between the arm links, which act as disturbance loads on the actuators. This causes significant tracking errors. Asada and Youcef-Toumi have presented an alternative design [1] based on a parallel linkage mechanism in which direct drive is achieved for the arm's shoulder and elbow joints using two stationary motors. Apart from making the moving parts of the arm lighter, they have shown that by selecting a certain mass distribution within the arm, invariant and decoupled reflected motor inertias may be realised.

Referring to Fig. 1, the mass distribution should be such that:

$$\frac{m_4}{m_3} = \frac{l_2 \cdot g_3}{l_1 \cdot g_4} \tag{1}$$

Since:

$$\begin{bmatrix} T_1 \\ T_2 \end{bmatrix} = \begin{bmatrix} H_{11} & H_{12} \\ H_{21} & H_{22} \end{bmatrix} \begin{bmatrix} \ddot{q}_1 \\ \ddot{q}_2 \end{bmatrix} + \begin{bmatrix} -K_1 & 0 \\ 0 & K_2 \end{bmatrix} \begin{bmatrix} \dot{q}_2^2 \\ \dot{q}_1^2 \end{bmatrix} \tag{2}$$
$$\tag{3}$$

Where:

$$H_{11} = I_1 + m_1 g_1^2 + I_3 + m_3 g_3^2 + m_4 l_1^2 \tag{4}$$
$$H_{22} = I_2 + m_2 g_2^2 + I_4 + m_4 g_4^2 + m_3 l_2^2 \tag{5}$$
$$H_{12} = H_{21} = (m_3 l_2 g_3 - m_4 l_1 g_4) \cdot \cos(q_2 - q_1) \tag{6}$$
$$K_1 = K_2 = (m_3 l_2 g_3 - m_4 l_1 g_4) \cdot \sin(q_2 - q_1) \tag{7}$$

When the parameters of the arms are chosen according to equation (1), the motors' torque equations simplify to:

$$T_i = H_{ii} \ddot{q}_i + T_{gi} \tag{8}$$

Where q_i and T_i are the angle and torque respectively, of the ith motor.
When the arm is mounted horizontally, the gravity torque T_{gi} is eliminated.

3.2   Open loop dynamics

Fig. 2 shows a block diagram model for open loop dynamics of a single joint consisting of a permanant magnet d.c. servo motor driven via a standard servo amplifier which contains an analogue current feed-back loop. This model is

valid as long as the parameters of the arm are chosen according to equation (1). The linear model assumes that the input U_d never causes saturation of the amplifier current rating or the motor torque rating. Normally, the trajectory generating software of the robot controller ensures the validity of this assumption [2].

It is emphasized that unlike the more conventional d.c. servomotors, the high performance motors used in DD arms have an inductance of the windings, L, too large to be neglected. This increases the order of the system model used in the analysis to two. At the same time the mathematical representation of the system can be slightly simplified by assuming that friction is negligible, which is reasonable considering the abscence of gearing.

From the block diagram, the open loop transfer function is given by:

$$G(s) = \dot{\theta}(s)/U_d(s) = \frac{K_c K_e K_a K_t}{JLs^2 + J(R+K_i)s + K_t K_b} \tag{9}$$

If K_i is sufficiently high so that $J(K_i+R)^2 >> 4LK_b K_t$ [3] then the two poles of $G(s)$ are given by:

$$P_1 = \frac{-K_b K_t}{(K_i+R)J} \quad ; \quad P_2 = -\frac{K_i+R}{L} \tag{10}$$

Then:

$$G(s) = \frac{K_c K_e K_a/K_b}{(\tau_e s + 1)(\tau_m s + 1)} \tag{11}$$

where the electrical time constant τ_e, and the mechaical time constant τ_m are given by

$$\tau_e = L/(K_i+R) \quad ; \tau_m = (K_i+R)J/(K_b K_t) \tag{12}$$

The mechanical time constant is directly related to the load inertia. Since there is no gearing it tends to be much larger than the electrical time constant, and therefore dominant. Further, it can be seen that the effect of current feedback is to reduce τ_e even more and to increase τ_m. Hence if K_i is sufficiently high τ_e can be neglected and the system model can be reduced to:

$$G(s) = \frac{K_{ii}}{\tau_m s + 1} \tag{13}$$

where $K_{ii} = K_c K_e K_a/K_b$ is the steady state open loop gain of the drive system. Note that this simplification is done at the expense of increasing the mechanical time constant. A block diagram of the position control system is given in Fig.(3).

3.3   Force control

Due to the absence of gearing (and the associated backlash and friction) in direct drive arms it was thought that by measuring the current in the armature, the torque exerted by each motor could be calculated. Consequently, by knowing

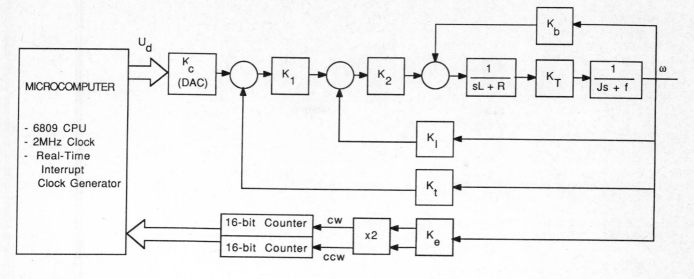

Kc : DAC Gain (V/BLU)　　　　　J　: Motor + Load Inertia (Nm/rad/s²)
K₁ : Control Amp. Gain (V/V)　　Kb : Motor e.m.f. Constant (V/rad/s)
K₂ : Drive Amp. Gain (V/V)　　　Kt : Tacho gain (V/rad/s)
K_T : Motor Torque Constant (Nm/A)　Ke : Encoder Gain (BLU/rad)
R　: Motor Arm. Resistence (Ω)　K_I : Current Feedback Gain (V/A)
L　: Motor Arm. Inductance (H)　f　: Viscous Friction Coefft. (Nm/rad/s)

Fig 3　　Control system layout of a single joint

the configuration of the arm, the force exerted at the end of the arm, F, can be determined from the motor torques, T, and by using the Jacobian matrix:

$$T = J^T F \qquad (14)$$

where J is the Jacobian matrix associated with the coordinate transformation from joint coordinates to the cartesian coordintes of the arm tip. When the arm has two degrees of freedom equation (14) becomes:

$$\begin{bmatrix} T_1 \\ T_2 \end{bmatrix} = \begin{bmatrix} l_1.sin(\theta_1) & -l_1.cos(\theta_1) \\ -l_4.sin(\theta_2) & l_4.cos(\theta_2) \end{bmatrix} \begin{bmatrix} F_x \\ F_y \end{bmatrix} \qquad (15)$$

Experiments using force transducers were carried out to verify the validity of the above statement in conjunction with the direct drive arm at Imperial College. The experimental results obtained showed that there is a linear relationship between the exerted torque by each joint and the numerical value on the corresponding Digital to Analogue Converter (DAC). The sensititvity of the force measurements was very good owing to the abscence of gears. Furthermore, it was established that by varying the microprocessor gain of the control system the mechanical stiffness of the arm can be changed on line.

3.4　Basic stiffness formulation

The rate at which forces and torques acting on the hand increase as it is deflected from a nominal position is termed stiffness. The basic stiffness formulation follows from a generalisation of the linear spring relationship, $f = K_{ii}.dx$, to an n-dimensional matrix expression:

$$F = K.dX \qquad (16)$$

where dX is a generalised displacement from a nominally commanded position ,X_0, of the hand, and the stiffness matrix in the 2 degree of freedom case is:

$$K = \begin{bmatrix} K_{11} & K_{12} \\ K_{21} & K_{22} \end{bmatrix} \qquad (17)$$

In our case:　$K_{12} = K_{21} = 0$

$$\theta_i - \theta_0$$

Defining $d\theta = \theta_0 - \theta_i$ as the difference between the actual joint angles and the nominally commanded joint angles:

$$dX = J.d\theta \qquad (18)$$

Combining eqs. (14),(16) and (18) the expression for joint torques necessary to make the arm behave as an n-dimensional spring in cartesian space becomes:

$$T = J.K.J^T .d\theta \qquad (19)$$

$K_\theta = J.K.J^T$ is called the stiffness matrix of the arm. Thus by varying the software gain of the controller, and therefore the stiffness matrix K, the stiffness of the arm can be controlled from the software.

The stiffness control method described here provides an intuitive format for simultaneous motion and force command. The method allows the programmer to think in terms of the stiffness at the gripper in a particular direction. For example, if the arm is expected to meet some physical constraints, then the stiffness matrix is adjusted for low stiffness in the direction of the constraint, while at the same time the stiffness in all other directions can be kept high. In this way the contact forces can be controlled.

3.5　Control model

The torque applied at the ith joint is given by the expresssion

$$T_i = T_{c,i} + dT_i + K_{v,i} J_{ii}.d\dot\theta_i + V_{0,i}.sgn(\dot\theta_i) + T_{gi} \qquad (20)$$

where $T_{c,i}$ =commanded torque, i^{th} joint , given by

$$T_c = K_\theta.d\theta + T_a \qquad (21)$$
$$T_a = J^T F_a \qquad (22)$$

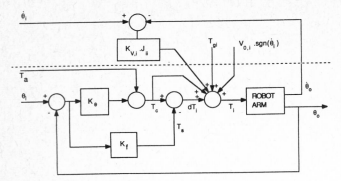

Fig 4   Force control block diagram

F_a is the applied controlled force by the arm. dT_i is torque error in the ith joint, given by:

$$dT_i = T_c - T_s \qquad (23)$$

T_s is the sensed torque ,given by:

$$T_s = K_f . d\theta_i \qquad (24)$$

The last two terms of eq(20) are due to friction torque and gravity loading respectively.In the case of the direct drive arm mounted horizontally ,both terms can be eliminated from eq(20).

There are situations when the arm is out of contact with the environment , the inclusion of the third term on the right hand side of eq(20), ensures that there is velocity dependent damping and hence large impact forces are avoided. A block diagram of the stiffness control system is given in Fig. 4. This system (excluding the loop related to the velocity damping term) is implemeted in assembly language on a 6809 Motorola microcomputer.The sampling period is 4 ms. A background job is run in a high level language, updating the values of J and K_θ.

As a result, a force sensing system has been developed without using any extra hardware. The need for complicated and expensive force sensors has thus been eliminated.

4   ADAPTIVE POSITIONAL CONTROL

From the control point of view, all robotic arms are troublesome systems since their serial link structure makes them highly non-linear while the dynamic interactions which normally exist between the arm links act as disturbance loads on the actuators.This can cause significant tracking errors, especially during high-speed motion, but in a conventionally designed arm the high gearing ratios act to mask the arm non-linearities, as sensed by the actuators, and make the whole system much more manageable. In a direct drive arm such masking by the gears does not exist.

As mentioned above, the dynamic problems can be largely overcome by careful design of the mechanics. The design presented by Asada and Youcef-Toumi [1] achieves invariant and decoupled reflected motor inertias for the arm's shoulder and elbow joints by selecting a certain mass distribution within the arm. It must be stressed, however, that the mechanical couplings and non- linearities are eliminated only for a particular weight of the payload.

The probem of changing dynamics therefore still exists to some extent and if the arm is to move accurately and at high speed then the controller must adapt accordingly. It has been suggested [3] that on-line modelling of the arm should be employed to calculate the dynamics. However, setting-up of such a model is highly arm-specific and prone to errors while the calculations are very intensive. Thus here we propose on-line *estimation* of the system dynamics to provide the controller with the necessary information.

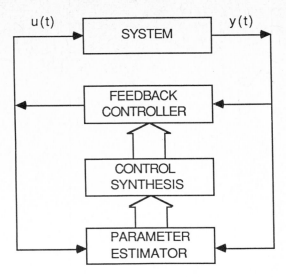

Fig 5   Self-tuning controller structure

The structure of the controller is depicted in Figure 5. It operates on the individual joint level. The estimator monitors the sequence of input and output signals in an attempt to establish the parameters of the joints' transfer functions and the most recent updates of the parameter estimates are used to construct the suitable control law.

The transfer function of a single joint can be represented in discrete form as follows:

$$A(z^{-1})y(t) = z^{-k}.B(z^{-1}).u(t) + C(z^{-1}).e(t) \qquad (25)$$

where $\{y(t)\}$is the sequence of output signals (joint position) and $\{u(t)\}$ is the sequence of input signals (motor voltages). The polynomials in the difference equation are defined as:

$$\begin{aligned} A(z^{-1}) &= 1 + a_1.z^{-1} + a_2.z^{-2} + ... + a_m z^{-m} \\ B(z^{-1}) &= b_0 + b_1.z^{-1} + b_2.z^{-2} + ... + b_n z^{-n} \\ C(z^{-1}) &= 1 + c_1.z^{-1} + c_2.z^{-2} + ... + c_p z^{-p} \end{aligned} \qquad (26)$$

z^{-1} is interpreted as the backward shift operator such that:

$$z^{-1}u(t) = u(t-1) \qquad (27)$$

The term $C(z^{-1}).e(t)$ represents external noise disturbance, whose parameters can also be estimated, and in this context it can be used to represent the dynamic coupling between the joints.

The estimator employs a Least Squares fitting technique to establish the coefficients of the polynomials. To do this the discrete system model of (25) can be rewritten in a more suitable form as:

$$y(t) = \phi^T(t).\theta + \varepsilon(t) \qquad (28)$$

where θ is the system parameter vector :

$$\theta^T = [-a_1, -a_2, ... , -a_n, b_0, b_1, ... , b_m] \qquad (29)$$

$\phi(t)$ is the vector of past observations:

$$\phi^T(t) = [y(t-1), ... , y(t-n), u(t-k), u(t-k-1), ... , u(t-k-m)] \qquad (30)$$

and $\varepsilon(t)$ is the system noise.

If N input-output pairs are recorded then equation (28) can be used to link all of the measurements as:

$$Y(N) = \Phi^T(N) \Theta + \varepsilon \qquad (31)$$

where:
$$\Phi^T(N) = [\phi(N), ... , \phi(1)]$$
$$Y^T(N) = [y(N), ... , y(1)]$$

Let us denote the estimates of the system parameters by $\hat{\theta}$. In order to evaluate θ we define the cost function

$$V_N(\hat{\theta}) = ||Y(N) - \Phi^T(N)\hat{\theta}||^2$$

$$= \frac{1}{N} \sum_{t=1}^{N} [y(t) - \phi^T(t)\hat{\theta}]^2 \qquad (32)$$

Clearly, the cost function is minimised when $\hat{\theta} = \theta$, which enables us to obtain $\hat{\theta}$ analytically. Differentiating $V(\hat{\theta})$ with respect to $\hat{\theta}$:

$$grad\ V(\hat{\theta}) = -2\ \Phi(N).(Y(N) - \Phi^T(N)\hat{\theta}) \qquad (33)$$

and at minimum $grad\ V(\hat{\theta}) = 0$, therefore:

$$\hat{\theta} = (\Phi(N).\Phi^T(N))^{-1}\Phi(N).Y(N) \qquad (34)$$

or equivalently:

$$\hat{\theta} = \sum_{t=1}^{N} (\phi(t).\phi^T(t))^{-1} \sum_{t=1}^{N} \phi(t).y(t) \qquad (35)$$

For on-line implementation eq.(35) is used in a recursive form which may be summarised as follows:

$$\hat{\theta}(t) = \hat{\theta}(t-1) + K(t).[y(t) - \phi^T(t).\hat{\theta}(t-1)] \qquad (36a)$$

$$K(t) = \frac{P(t-1).\phi(t)}{\lambda(t)/\alpha_t + \phi^T(t).P(t-1).\phi(t)}$$

$$= \alpha_t P(t-1).\phi(t) \qquad (36b)$$

$$P(t) = \frac{1}{\lambda(t)} \left[P(t-1) - \frac{P(t-1).\phi(t).\ \phi^T(t).P(t-1)}{\lambda(t)/\alpha_t + \phi^T(t).P(t-1).\phi(t)} \right] \qquad (36c)$$

The parameter estimates can be used to construct the control law. Assuming that the feedback control is applied in standard form we can write:

$$F(z^{-1}).y(t) + G(z^{-1}).u(t) = 0 \qquad (37)$$

where $g_0 = 1$, resulting in the closed loop system to be given from (35) and (37) by:

$$(A.F + z^{-k}.B.G)y(t) = F.Ce(t) \qquad (38)$$

where $(A.F + z^{-k}.B.G)$ forms the characteristic equation of the closed loop system.

Let the desired closed loop performance of the system be given by the polynomial

$$T(z^{-1}) = t_0 + t_1.z^{-1} + ... + t_{nt}.z^{-nt} \qquad (39)$$

which determines the poles of the closed loop system. If $F(z^{-1})$ and $G(z^{-1})$ are chosen so that the following relationship is satisfied:

$$A.F + z^{-k}.B.G = C.T \qquad (40)$$

then the closed loop system transfer function becomes:

$$y(t) = \frac{F(z^{-1})}{T(z^{-1})} e(t) \qquad (41)$$

The pole placement controller of this types allows for optimal control strategy. If the closed loop poles are placed at the origin, this effectively means that the controller is trying to reduce the output error to zero in minimum time, i.e. within one sample interval. Similarly, control can be readily detuned under software by moving the closed loop poles away from the origin.

The controller of this type has already been implemented on a conventional type of anthropomorphic robot and was shown to perform very well. It is planned to implement it on the new DD arm as soon as it becomes available.

5 CONCLUSION

Direct drive arms show a very good potential for constructing a reliable, low-cost, high-performance robotic system for automated assembly. In such a system as much of the control problems as possible should be solved in software. The proposed control strategies tie in very well with this idea since they allow active compliance and dynamic compensation, respectively, without any need for specialised sensors. Morover, they allow system performance characteristics to be changed on-line under program control to suit the task which is being executed. Both control methods have been verified experimentally.

REFERENCES

(1)   ASADA,H. and YOUCEF-TOUMI, K. "Analysis and design of a direct drive arm with a Five Bar Link parallel drive mechanism." ASME journal of dynamics, measurement, and control, vol. 106, no. 3, 1984, pp. 225-230

(2)   HOLLERBACH,J.M.,"Dynamic scaling of manipulator trajectories",AI Lab,MIT, AI Memo no.700, Jan 1983.

(3)   PAK, H.A. and TURNER, P.J."Optimal tracking controller design for invariant dynamics direct drive arms".Internal report, Centre for Robotics and Automated Systems , Imperial College of Science and Technology, London, England, 1986.

(4)   RISTIC, M, "Dynamic analysis and control of anthropomorphic robots, Ph.D. Thesis, University of London, May 1986.

C379/86

The redesign of a manufacturing business

M W DALES, BSc, CEng, FIMechE, MBIM and P JOHNSON, BSc, PhD, CEng, MIM, MIProdE
Lucas Systems Engineering Projects, Solihull, West Midlands

SYNOPSIS

The paper is based on experience in Lucas working from the need to achieve competitive levels of performance in a manufacturing business which has been subject to incremental change over time. It describes the process of re-designing such a manufacturing unit starting from clear business objectives, marketing strategy and product strategy; it is a systems design approach leading to the re-structuring of the business into modules followed by simplification of the organisation and control systems. Methods of tackling this re-structuring are discussed with examples of their use. The suitability of the design is assessed both for the steady state and dynamic conditions which are likely to be encountered. The application of MRP II and Kanban are discussed, together with composite or hybrid control systems which draw on appropriate elements of both types of control. The impact of the re-design on overhead service functions and cost structure is mentioned together with cost benefit analysis. Examples of actual achievements in Lucas are illustrated. Some comments are made regarding the economic implications to the UK of moving to short lead time manufacture.

INTRODUCTION

Most manufacturing businesses evolve over a period of time, during which they are subject to countless incremental changes. These changes result from new product introductions, from the addition of many minor options and from changes in product specification. They also result from the ways in which manufacture is carried out' and organised, such as the introduction of new machine tools or processing plant, new processing methods or changes in working practice. As well as all these relatively small, incremental changes which occur, there are of course from time to time significant step changes resulting perhaps from introducing complete new product designs or maybe acquiring another related business and consolidating its manufacture into the home base.

A further complexity results from the migration of products through their life cycle from prototypes through original equipment manufacture, to product support serving an after market. This often results in the long-standing supply of products or perhaps just components in small volumes at irregular times, using plant and tooling which is designed for high volume manufacture. At the same time the business may be seeking to achieve changing levels of performance, particularly in the quality arena.

In order to cope with these complexities, and others, control systems are overlaid upon the factory, its materials and its people. Not only are these complex at the start, because of the nature of the control task, but they naturally tend to increase in complexity as those who exercise the control functions, seek evermore information to unravel the mysteries of their task. The control tasks may themselves be aggravated by the organisation structure if it is a highly functional one, with large numbers of different specialist functions or departments contributing to the whole because each additional function potentially creates more communications and co-ordination requirements on top of the fundamental ones that existed in the first place.

Whilst this may, at first glance, appear to be a caricature, there is no doubt that there are many manufacturing businesses which have experienced all of these problems and indeed others. This being the case, it is little wonder that complexity increases over time and levels of performance, unless they are sustained by continuous volume growth, tend to decline. If such a situation is encountered, what is actually required, is a methodology for first of all understanding the fundamental issues and secondly, unscrambling the complexity, so that an effective arrangement can be established for the design and manufacture of a range of products in a competitive manner.

This paper outlines the methodology developed and employed by Lucas for re-structuring and simplifying a manufacturing business, in other words for the re-design of a manufacturing business. What is described is a practical and proven set of methodologies, which if used with skill and discretion are effective over a wide range of different businesses. These design methodologies are developed from the five basic stages of designing manufacturing systems (1) which are set out in Figure 1. What must be clear however, is that this, like any aspect of manufacturing systems engineering, is not a methodology characterised by working to a manual and filling in pro-formas, then turning the handle in order to get an answer; it is in fact a very powerful model which needs to be drawn on with care and skill, so that it is appropriately directed in different circumstances.

Competitive Performance Targets

The starting point in the re-design process must clearly be establishing the measures of competitive performance (2). This may not be a simple task but it is clearly essential for any business which seeks both to protect its position in the market place and indeed improve that position. In today's competitive environment, a small improvement is unlikely to be sufficient. The starting point is therefore setting these competitive objectives after the necessary intelligence gathering and analysis has been completed. This is a specialist subject area in its own right, which is not included in the scope of this paper. It is assumed that the business can make a

reasonable and effective estimate of appropriate competitive targets for such business measures as sales per employee, stock turn ratio, lead time, selling price, product cost and measures of quality. These are of course typical performance measures and in some circumstances others should be included. This therefore, is the starting point, as indicated in the left-hand block of Figure 2 which is a block diagram illustrating the first stages of the re-design process.

Marketing and Financial Analysis

The next step is to carry out marketing analysis and financial analysis. The order in which these are tackled and the balance between them depends on the nature of problems being addressed. For example, many consultancies will naturally approach from a financial analysis point of view, whereas others will approach from a marketing point of view. In our experience, detailed financial analysis is probably only necessary at this stage if in a multi-product factory the existing financial information systems are extremely misleading or in part non-existent; it is the intention after all to re-design the business and this will take away the foundation stones on which the existing costing systems depend, which means there is limited value in investing effort at this stage to the financial issues.

The market analysis can be done in a variety of ways but should include a SWOT analysis with a situation review and detailed sales and market analysis. SWOT analysis is a technique for assessing strengths, weaknesses, opportunities and threats, which pertain to the business; hence the acronym SWOT. It is an internal and external examination of the relationship of the business with its market and environment. It provides a disciplined approach for matching strengths against opportunities and protecting weaknesses from threats. The situation review which could be described in a number of other ways will identify the market place, market share and competitive position as well as trends in the market place itself. The detailed market analysis considers product life cycles and future demand levels.

The marketing function may be unaccustomed to providing a level of detail which is suitable for a proper manufacturing systems re-design. It is therefore necessary also to examine detailed sales history in order to understand the fine details from volume, variety and frequency fluctuations. These of course have to be corrected for any special historical factors whether they be one-off incidents or simply the difference between demand and sales.

Marketing and Product Strategies

The remaining element at this stage is the market and product strategy. These need to be statements of objectives and strategies both in marketing and product development. It is important that these are clear and agreed statements, so that an appropriate manufacturing strategy can be developed to properly complement and support the marketing, product and business strategies. It is of absolutely no use having a marketing strategy which sets out to sell quality and reliability of service in a variety market, if the manufacturing strategy focuses its efforts on cost with limited regard to and different priorities for the other dimensions (3).

What we have at this stage therefore, are detailed planning guidelines for volume and variety over time, together with a clear statement of what the manufacturing system is required to do at a strategic level. There are bound to be uncertainties about these guidelines but no matter; they are better than blind guesses and if the appropriate responsible people are required to validate and sign them, including ranges of uncertainty, the design work can proceed.

Manufacturing Data

Having established firm marketing foundations it is then necessary to start collecting manufacturing data that is appropriate to the re-design task. Before collecting data, it is important to have an understanding of the uses to which that data will be put - data selection before collection is the catch phrase. This underscores the need to follow the business re-design methodology not in a simplistic way but to use it with intelligence and discretion.

For instance, it may be important to select parts on the basis of weight or on the basis of the materials from which they are made, in which case it is important at the data collection stage to identify the weight of the products or components, or materials from which they are made. In addition, it may be important at this stage to begin identifying the characteristics of the individual products and components as the information is collected and subsequently stored. In this way the data collection stage is a constructive process and can start to help in preliminary identification of means of simplification.

As well as collecting data in this fashion, it is also important to store the data in a way which will make it traceable and easy to retrieve and manipulate. It should therefore be entered into computer data files so that analysis can be undertaken quickly by the computer. In fact many of the techniques which are used to re-design a manufacturing organisation such as production flow analysis become almost impossible once a modest complexity level has been exceeded unless handled by computer. However, in all but the most extreme cases this can be the ubiquitous personal computer with propriety software.

As well as the collection of information on the products to be produced, data needs to be collected on the manufacturing processes employed and the flow of work within the factory. This is done by use of flow charts. It is important once again to understand the use to which this flow chart information will be put so that priorities can be established. As well as charting the actual manufacturing routes, which may differ from the planned ones, other factors which need considering include distance travelled, transfer/process batch sizes and inventory levels.

At this stage the identification of "no-value-added" activities will become apparent and can be indicated on the flow charts. These include inspection, storage, transport and rectification activities and may include some actual operations as well. No-value-added is used here in the literal sense, not the accounting sense, and may include some operations.

Equally, it is necessary to collect data on the capacity, speed, reliability, process capability and changeover times of the machines and processes. All of this data will be required for the establishment of manufacturing cells or modules. For example, if we are looking to establish a manufacturing cell to produce high variety with short lead times, then the most important factor could well be the amount of time required to change the machines over. Equally if we decided that our manufacturing cells are to work Just In Time, perhaps producing a fairly standard range of products, but on a make daily, sell daily basis, then machine tool reliability will be just as important.

It is clearly vital to collect information about the quality performance; the amount of scrap which is produced, the re-work figures, the original equipment returns from customers and warranty performance. Considering these factors plus the cost of fault detection and obsolete stocks, together with no-value-added activities and opportunity costs of non conformance, quickly points the way to how significant performance improvement can be achieved.

Outlined above is just some of the information which may need to be collected, analysed and used in the re-design of a manufacturing system. It should now be clear that it is important to look ahead into the re-design process and to focus continually on significant information. For example, less data will be required if it is not the intention to re=structure into cells. However, our experience clearly indicates that reduced benefits will be obtained if this key step is omitted. One of the many results will be an action plan for "cleaning up" the data to make it complete, accurate and singular.

Analysis and Sorting into Families

Techniques are available for sorting data into useful patterns. The first and simplest is Pareto Analysis to prioritise parameters. This technique allows concentration of effort on those parameters which can provide the biggest return of time and effort employed and it should be used all the time.

The key work, however, is sorting into families employing such factors as batch size and frequency, materials used, physical attributes or size of the products. In this way families can be formed which will provide the basis of the module designs. The trick is to find the family relationships which allow definitions of modules so that groups of products or components can be taken through a series of processes with simple material flow patterns. The result is a collection of modules, some in parallel, others in series, with crisp boundaries having the minimum material and information flows across them. Before and after material flow routes from a real example are illustrated in Figure 4.

Undoubtedly there will be some "problem children" which do not fit comfortably. In these few cases the module design should not be prejudiced until the manufacturing routes have been validated or the marketing function has been challenged to see if it really is worth making these misfits or a buy-out option has been explored. It is at this point that the importance of integrated design and manufacture start to become evident and some misfits can be accommodated after design changes.

This is essentially a group technology approach but it is a pragmatic one which may result in hybrid solutions. When considering analysis methods, production flow analysis (4) is preferred to classification and coding in most cases because it is the method of manufacture that is important not the product itself. There are also important options to consider in multi product factories as to whether the focus of grouping is purely on finished products or whether some or all component manufacture is de-coupled from the finished products and separately grouped. This will have a significant effect on control systems as well as the architecture of the factory. Equally a pure jobbing shop will be better with process orientated cells. Many of the concepts applied to cells, like ownership, measures of performance, clear definitions of boundaries and minimum dependence on outside support aplly whether or not they are product (group technology) based.

Buy-out options should be encouraged for minor and non-strategic components but must not be considered except as a temporary measure for strategic components and processes which contain distinctive competence and can provide competitive advantage.

The result is a new modular structure to the factory which provides a simple foundation on which all else can be built with relative ease provided there is attention to detail. It represents a major milestone in the design methodology.

Steady State Design

Having defined the modules the next step is to carry out a steady state design, as indicated in Figure 3. Steady state conditions comprise the most likely volume and mix of products together with the average performance of machines, labour, suppliers. It must consider not only the work stations and the modules but also how they fit together.

At this stage capacity must be matched against demand and the problems and bottlenecks begin to show. Consequently existing approaches to manufacture should be challenged and plans made to reduce or eliminate the causes of problems, not their symptoms. Examples include reduction of changeover time, which is often a surprisingly cheap and feasible thing to do, and overhauling machines to re-establish process capability. All the necessary machine and material resources for operating the modules need quantifying in detail. Material must not be regarded as a free issue resource.

A vital part of the steady state design is job and organisation design. People are an integral part of the manufacturing system and it is usually impossible to gain significant performance improvement unless the systems approach extends to job design. Whilst many may be uncomfortable at the thought of implementing such changes, they are essential. This must result in a simple organisation structure with few levels and few functions; it must be a highly flexible arrangement which will widen and increase the skill requirements of most jobs; it must pull into the modules as many indirect support activities as can be practical and economic. Only then can the information flows across the module boundaries be minimised, ownership and accountability be established and headcount reductions be made in the staff and indirect areas.

To ensure success, care must be taken to identify the skills required for each job, to audit the skills of existing employees at all levels, and to establish the nature and volume of training needs. Practical experience suggests the estimated quantity of training should then be increased by perhaps a factor of two! Methods of selection for the new jobs will also be needed along with a policy for those who don't make the grade. Whilst the benefits of competitive payroll productivity are high, the failure risk is also large when the task is measured against vested interest and resistance to change.

After re-structuring job functions and responsibilities inside the modules the next step is to specify the reduced service levels which are required from outside the manufacturing modules. This provides the foundation firstly for reduction in numbers and secondly for another re-design task, namely the simplification and re-design of all the service and other non-manufacturing functions. Equally it is here that the organisational issues underlying integrated design and manufacture must be addressed. Many Lucas business units have now chosen to create a single Technical Function which is responsible for both product development and manufacturing systems development. These changes are excluded from the scope of the paper.

Dynamic Design

The steady state design must then be tested against likely change of variables. These may be a reflection of uncertainty about the planning guidelines derived from the marketing analysis or variation in performance of any aspect of the system, whether it be machine reliability, labour, productivity, or whatever.

Dynamic design can be carried out using simple paper calculations or spreadsheets; it can employ manual or computer simulation models. Even if using a computer model, Lucas experience demonstrates the benefits of building a manual model first to aid understanding and reduce the risk of the model containing unrealistic assumptions. The point, of course, is to use whatever tool is cost effective depending on the complexity and implications of the dynamics. The results will vary between minor adjustments of fixturing and plant layout, to the discovery of major pitfalls which dictate the need for re-appraising the design.

One of the more complex dynamics is sometimes found to be the transient condition during the change from the current situation to the new design. With new plants this is sometimes known as the run up period, but metamorphosing an old style factory into a new style is in many ways more complex.

Control Systems Design

Part of the analysis work includes examining the existing control systems in order to identify any weaknesses. Such weaknesses are likely to make themselves visible in the form of inadequate data interchange between the planning system in the business area and the execution system. The new system design is based on learning from any imperfections in the old system, taking into account changes to manufacturing techniques. Another weakness sometimes found is ineffective capacity planning either prior to or after acceptance of orders.

In Lucas advantage has been taken of simplified modular factory designs to fundamentally re-think how the business and manufacturing control systems should work. By the use of small modules the possibility of a module based work load acceptance routine is introduced using decision support systems based on small computers in the modules. Such tools can be made far more dynamic as a result of the smaller span of control they exercise with consequent reduction in data requirements when compared with a large planning system covering the site. The work has advanced to the stage of distinguishing between the types of system required for working in environments which range through a spectrum between pure JIT (Kanban) (5) which need very little support, to MRP based systems (6) for non-repetitive work with complex operations. The result is a family of hybrid, or composite control systems which allow migration from

one to another. For example, Figure 5 illustrates the concept of devolving appropriate control to the modules leaving a simplified material requirements planning system at business level.

It should not be inferred that Just In Time is purely about control systems. JIT is a much wider concept than this which has more to do with continuous improvement than any particular aspect of manufacturing, such as production/material flow control.

Figure 6 shows the two types of control sytem being developed and used within Lucas. The right hand half of the figure shows a module control system for modules suitable for Kanban. The only inputs required to this system are the demand to final assembly and the materials to goods inwards. The system is designed to pull from final assembly back to raw materials (and ultimately across suppliers' factories). This is an elegantly simple solution where manufacturers work to meet visible real demand and are encouraged to stop work if it is not required or defective in some way.

Whilst this system requries the majority of work to be made regularly a significant demand for irregularly made products or spares can be accommodated by injecting single cycle Kanban's into the appropriate part of the system.

The other half of the figure shows the use of a Lucas developed in-cell scheduler used to optimise the workload of a cell. This system is required where the majority of work consists of high variety low volume batches not made regularly. This requires dynamic user specified schedulers which are linked to an MRP I system which has been adapted to allow the passing of blocks of work to a module or cell.

In both systems a key issue is the use of three levels of workload planning and acceptance. One level looking some distance into the future owned by the business and two levels owned by the module. These latter two levels carry out module based order acceptance and subsequently scheduling of the work through the module.

The implications of using small module based computer systems in this way is that the number of functions required for the business planning system is significantly reduced. This has the advantage that the responsibility for maintaining the data up to date can more easily reside with those who own the data. Obviously this situation requires formalising and the design work has to include formal mechanisms to ensure that both systems run smoothly in tandem with accurate, singular and up to date data.

Whilst the design is aimed at the new style of operation, account may need to be taken of the transient phase during which the lead times in manufacturing will drastically reduce while the stock turns rise. It is only by the use of small highly dynamic shopfloor based systems that the manufacturing system can cope with this stage.

The control system needs to provide meaningful performance measures for modules as well as the business. These will include output by type (not just volume), stock turn ratio, leadtime and quality achievement. It may need to provide support tools such as quality diagnostics, maintenance aids and tool tracking aids.

Costing

Having arrived at a statement of the machines required, the plant layout, the manning levels necessary, he format of any services required and the control system, it is possible to begin to calculate lead times through manufacturing and the new stock levels required and hence costs.

When building up the cost estimate it is vital to distinguish accurately between costs which are genuinely attributable to the modules and those which belong to the factory and business. All services used by the module should be charged at true cost rates so that providers of the service are obliged to perform competitively and have their own performance targets. This is a bottom up approach which accounts for the real use of resources and allocates overheads in a meaningful way. It is quite different from standard costs plus overheads apportioned in an arbitrary way (7).

Make versus Buy Analysis

Upon completion of the basic design, and understanding the more competitive cost structure which results, it is then feasible to undertake a 'make-versus-buy' analysis. Although both strategic decisions and first pass estimates can be made earlier, it is clearly important not to make final decisions until a proper understanding of the cost structure of the re-designed business has been obtained. Make versus buy decisions maybe incremental in dealing with relatively small numbers of components which simply are an irritation and do not fit into the modular design. Or they may be of a larger scale. This perhaps could be equivalent to considering whether to buy a complete range of components or a particular process, that is effectively to exclude one particular module from the design and rely upon external suppliers in its place.

Care is required in either case to ensure that the financial analysis is a real one, based upon the actual impact on commitment and cost in and out of the business resulting from the decisions including all allocations for overhead and service areas as well as the direct manufacturing activities. These implications may be quite different for the incremental component by component make versus buy analysis as compared with a larger scale analysis.

Supplier Development

It should now be evident that whilst there are enormous benefits obtainable from re-designing what goes on inside the business, the overall performance levels can be further improved if a similar approach is taken further back up the supply chain (8). The first step of course in this process is a rationalisation programme so that the task is manageable. This is not to say that supplies are necessarily single source but that vast proliferation of suppliers, materials, bought out components and sub-assemblies is clearly not consistent with close effective working relationships and high levels of competitive performance. It is actually a supplier integration process which seeks to establish effective supply relationships between modules inside the factory and modules outside the factory. To do this requires two things. The first is fairly obvious, and is to provide support, expertise and perhaps even resources to help key suppliers go through similar processes to the ones which you yourself have been or will be going through. The second is perhaps less obvious and that is to look inwardly at your own attitudes towards suppliers, your relationships with them and how you control them.

One of the results in some Lucas businesses has been the creation of a supplies module which is responsible for supplier development, purchasing, procurement and storage of raw materials and components. This fits well with simplified control systems and provides a clean interface between the in-house manufacturing modules and the supplier network.

Planning Implementation

In order to receive approval to put the new system in place it is necessary to plan in some detail the method of implementation. It is only by doing this in detail that the true cost of the moves necessary can be ascertained. In producing the plan, account has to be taken of customer demand profiles and shut down periods in order to be able to satisfy both the market requirements and achieve the changes in the shortest possible time at the least cost. It may be feasible to carry out some preliminary moves quite quickly whilst later moves may have to wait for factory shut down periods.

The control systems usually require more care even than the machine moves and modifications. Migrating a business from one set of control systems to another from material flow control up to business planning is far from simple.

The greatest task of all however, is planning the people system changes. This involves designing the training material and establishing appropriate training plans for each and every individual as they will all be doing at least partially new jobs, demanding new knowledge and skills. More exacting still is the task of instilling new attitudes and disciplines into everybody which strike at the root of established practice. This requires

considerable education, training time and resources; it requires firm leadership with values being set by leaders and much of the training being done by the leaders themselves. Although every means should be used including conventional training, hands on training and distance learning, it is no good passing ownership of the education and training task out to someone else.

There are of course industrial relations issues wrapped up in this area. Apart from advising an approach which is both direct and professional, this paper is not the place to discuss such issues which are heavily dependent on local factors.

All the elements of the implementation plan must be integrated into a master plan with clear allocation of responsibilities, and effective command and control mechanisms.

Looking beyond the implementation itself it is also important to put in place mechanisms for continuous improvement so that the new design can flex and adapt to changing requirements and competitive pressures.

Financial Assessment

All this work is not of course done as an intellectual exercise, but in the pursuit of those competitive business objectives which mark the starting point. The closing activities in the re-design process are therefore financial. It is necessary to make detailed estimates of costs and sales income associated with the new design, as a first step. The next step is to bridge the time period from the current position to the fully implemented and established position, which will be a dynamic and complex period. It includes examining all the cash flows for capital expenditure, reconstruction, extraordinary expenditure, inventory movements and write-offs, as well as income. Only if this is thorough and detailed with a properly integrated set of trading accounts, project estimates, balance sheets and cash flow profile, can a real financial assessment be made with due regard to appropriate pay-backs, return on capital and discounted cash flow (9).

It is vital in this financial assessment process to recognise that what is being designed is a fundamentally different manufacturing business with different performance levels and proper regard being paid to all the benefits of this, some of which may seem to be a little intangible.

What should be emphasised at this point is that it is not necessarily a financially difficult project to implement. Clearly there are fundamental issues to attend to such as the replacement of decrepit equipment, which is not process capable, and this will require investment. Clearly, however, it has a payback in reduced cost of quality and is necessary regardless of undertaking the re-design project. Experience indicates that for a financial investment which might be between 5% and 15% of annual turnover, a typical high variety, repetitive manufacturing business can be re-structured and significant benefits and good pay-backs obtained. Indeed in many circumstances, with the exception of small disturbances it can even be a cash neutral project with much of the early funding coming out of the inventory reductions and some of the later funding coming out of reductions in payroll.

The critical area however is not the financial resource but the people resource. The critical path will inevitably revolve around the number and the quality of skilled people necessary to undertake the training, to fill the new job structures and to complete all the tasks.

Achieved Results

This general approach has been adopted over the last 2 years in Lucas for re-structuring a large number of business units. The earliest of these projects have now progressed well into the

implementation phase with measurable achievements along the way (10). At a small scale, achievements include reductions in manufacturing lead times of automotive products by a factor of 10 times and more. At a larger scale, some large automotive business units are eighteen months into, and more than half way through a re-structuring plan which has resulted in stock reductions of 50%, reductions in defect levels of several times and productivity improvements of up to 20%. At the end of this journey the units will

be in a fully integrated, Just In Time mode of operation with key suppliers working JIT as well.

In the aerospace sector there are similar examples of performance improvements as well. These include a reduction by more than 50% of lead times in component manufacture and a doubling of effective use of direct labour.

All these examples are measures along the journey of improvement and are not from the end or final achievements. Interestingly, the motivation and commitment of those who have taken part in the change process has also increased significantly.

It is important to emphasise that these achievements have resulted from no new expenditures on sophisticated machine tools, CIM, robots or any other from of expensive technology. This is not to say that technology has no place when clearly it does. Investment in technology offers further benefits if it follows diligent application of methodologies which are strategically sound and systems based. Expenditure on technology in advance of attending to these fundamental issues is destined for disappointment and failure.

Economic Implications

As a final point, it is worth reflecting on the economic implications of this work. Whilst a rigorous economic analysis is not claimed, key questions arise.

Will international competitive pressures drive all UK manufacturing industry and its distribution channels into short lead time operations? What are the consequences in terms of productivity improvements and the release of labour for new employment or no employment? If there is a general move towards lead time reductions, what is the size of the once-only reduction in output as stock levels are reduced to satisfy orders at the "expense" of throughout and material procurement? Do current economic forecasting models recognise the impact of this output reduction on gross domestic product over the next few years?

It is notable that published treasury figures show that stock-output ratios for both manufacturers and distributors has been reducing since about 1980.

Causes to date are attributed to recession induced caution, in particular expectations of output, prices and interest rates, and also the effect of the 1984 abolition of tax relief for depreciation in value of stocks.

Even if some sectors of manufacturing are excluded, such as metals, chemicals/manmade fibres, food/drink/tobacco, the value of all stocks and work in progress of the remaining sectors was £21.6 billion (11) at the end of December 1984 (valued at 1980 prices). At the same date the stock value of the wholesale distributive trades was £9.5 billion. It seems conceivable that up to 50% of these stocks might be eliminated so that even if the process takes several years this could have a noticeable effect on gross domestic product during the transitional period. Alternatively, failure of manufacturing industry to grasp the nettle of such competitive forces will certainly have even more significant effects.

REFERENCES

1) 'The Design of Competitive Manufacturing Systems'
Author: J Parnaby
Proceedings on Fourth International Conference of Systems Engineering, Coventry Polytechnic, 1985

2) 'Where Lucas Sees the Light'
Publication: Management Today - 1985

3) 'Manufacturing Strategy'
Author: T Hill
Publisher: MacMillan Education Ltd

4) 'The Simplification of Material flow Systems'
Author: J L Burbidge
Publisher: Int. J. Prod. Res. 1982 Vol. 20 No 3, 339-347

5) 'Japanese Manufacturing Techniques'
Author: R J Schonberger
Publisher: Free Press 1983

6) 'A Comparison of Kanban and MRP Concepts for the Control of Repetitive Manufacturing Systems'
Authors: J W Rice, T Yoshi Kawa
Publication: Production and Inventory Management First Quarter 1982

7) 'Yesterday's Accounting Undermines Production'
Author: R S Kaplin
Publication: Harvard Business Review July/August 1984

8) 'Lucas Goes for Total Quality Control'
Author: J Bache et al
Publication: Your Business - January 1986

9) 'Total Project Management - Book 2- The Management of Finance'
Author: B Cox, M W Dale
Publisher: British Institute of Mangement

10) 'Systems Approach to Business Redesign'
Author: J Wood
Proceedings of the First Internationl Conference 9 April 1986 - Just in Time Manufacturing

11) 'Monthly Digest of Statistics'
Publication: CSO May 1986

1. INPUT — OUTPUT ANALYSIS working from Market and Product needs

2. STEADY STATE DESIGN using specifications from 1. above

3. DYNAMIC DESIGN for changes from operating levels in 2. above

4. Specification of DATA COLLECTION and INFORMATION FLOW functions

5. Definition of the separate CONTROL FUNCTIONS & CONTROL SYSTEM DESIGN

Fig 1   The five stages in manufacturing systems design
[J Parnaby, Fourth International Conference on Systems Engineering, Lanchester Polytechnic, 1985]

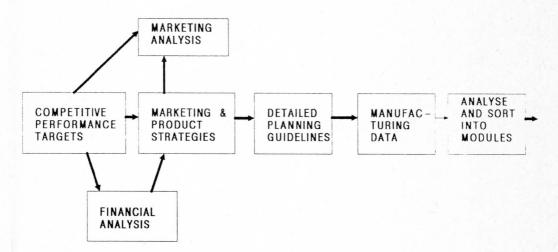

Fig 2   The redesign of a manufacturing business — part 1

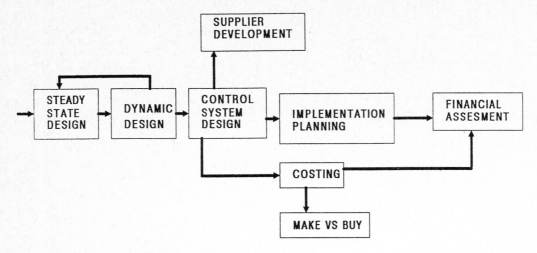

Fig 3   The redesign of a manufacturing business — part 2

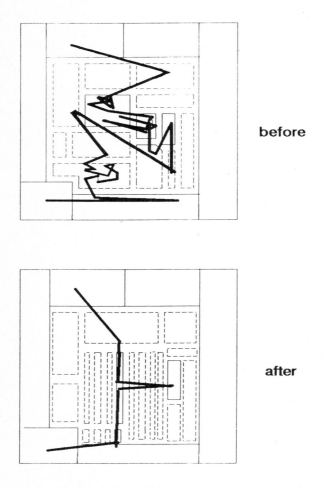

before

after

Fig 4   Material flow simplification

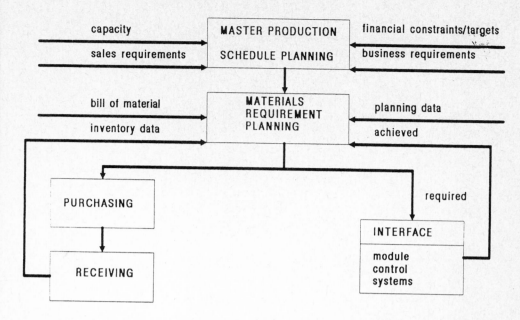

Fig 5 Simplified material requirements planning (MRP) based control system

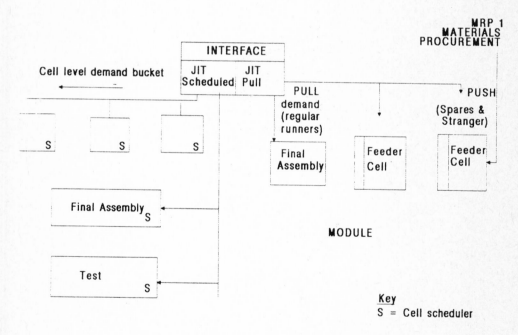

Key
S = Cell scheduler

Fig 6 Just in time (JIT) based module/cell control system

C380/86

ESPRIT flexible automated assembly cell— project aims and a progress report

J B MILLS
Westland plc, Yeovil, Somerset
A W WORSDELL
Normalair-Garrett Limited, Yeovil, Somerset

SYNOPSIS The context of the project is described with reference to a definition of 'Flexible Assembly'. The aims of the project are set out. Work-in-progress is related. Future work is described.

1   INTRODUCTION

The ESPRIT Directorate and the involved partners of ESPRIT Project S34 are committed to the dissemination of the results of our work. We therefore welcome this opportunity to present a statement of our aims and a work-in-progress report.

The term 'flexibility', in the context of an automated assembly system, is taken to have the following definitions:

- the ability of the system to be reversibly changed over to different tasks, involving different assemblies, sub-assemblies or component parts, whether planned or unplanned;

- the ability to produce a uniform output despite uncertainties in component parameters such as position, orientation, quality or tolerances;

- the ability to re-use system elements elsewhere when production ceases.

Systems capable of machining small batches of parts with rapid changeover to other batches as required are now a reality. By contrast, despite a great deal of research work, flexible assembly systems operating industrially are almost always targeted on a narrow range of products involving small detail changes in medium size batches.

Reasons for this lack of progress are: the limited performance of currently available robots in terms of speed, accuracy and repeatability; that, of all manufacturing areas, assembly is the one which requires the greatest application of human attributes such as vision, tactile sensing, force control, and intelligent reasoning; scheduling of work flow and full integration of hardware present severe problems; conventional cost justification techniques may not show flexible assembly systems in a good light; and, most products are currently designed for manual assembly and may require extensive re-design to render them suitable for automation.

It is the intention of this project to extend the existing level of flexibility by developing an automated assembly system capable of small batch, random production of a wider variety of products in an industrial situation.

2   THE APPLICATION ENVIRONMENT

Normalair-Garrett Ltd., part of the Westland Group, provides the basis for this development. NGL manufactures a wide range of mechanical and electro-mechanical assemblies, almost entirely for the aerospace industry. Components are typically of a volume less than a cube of side 0.5m, and of weight under 30kg. There is a very large product variety, and many are complex. Manufacture is to order, and usually in small batches, typically 20 or 25. Assembly is entirely manual, and in general one operator will assemble one complete batch. The nature of the applications demands very high quality levels, strictly controlled production conditions, and complete historical parts control to ensure traceability or origin.

Within the product range it is possible to identify 'families' of components which are naturally grouped together through having a range of common assembly processes. It is believed that a true flexible assembly system should address the production of a complete family, rather than small variants of one component.

To achieve such flexibility a number of approaches must be adopted simultaneously. The system will require offline programming to permit changes of product with minimal system downtime. Expert system techniques will probably be required to cope with work flow scheduling. Hardware and sensors must be integrated into a hierarchical control system with distributed processing and sophisticated communications. Mechanical hardware should be re-configurable to minimise costs associated with product-specific items. Sensors should be used to resolve uncertainties in the workplace. In particular vision, tactile and force sensing techniques are important. Provision must be made for incorporating error recovery techniques into the system. Finally, since it is impossible to fully automate assembly, especially where products have been designed for manual assembly, care should be taken to properly integrate humans into the system at all levels and all points of interaction.

3   THE PROJECT STRUCTURE

The project is aimed at the development of an operational demonstration flexible assembly system to be set up at a trial industrial site provided by Normalair-Garrett Ltd. The project is structured into various sub-areas, namely:

- system design and integration;

- development of enabling technologies, including gripper design, tactile and force sensing, robot control and off-line programming, vision sensing using optical image processing, and optical dimensional checking;

- role definition, being a study of the applicability of the technology to future manufacturing, including preliminary cost-effectiveness studies and study of design for manufacture aspects;

- research into the human factors of the cell with particular focus on its ergo-nomic aspects (including software ergo-nomics), on the design of jobs and the organisation of work within the cell, and on some of the wider organisational aspects of computer integrated manufacture;

- collaboration between psychologists and engineers to identify the extent to which human factors and technical con-siderations are congruent, to identify conflicts and trade-offs between them, and to optimise the overall performance of the system.

There are five project partners: Westland plc, operating through their subsidiary Normalair-Garrett Ltd; the Free University of Brussels (VUB); the Risø National Laboratory and Dantec Elektronik from Denmark; and the Social and Applied Psychology Unit of the University of Sheffield (operating with Medical Research Council funding) from the UK. See Fig. 1.

The partners' roles are as follows:-

Westland are project managers and are con-cerned with the specification and design of the overall system, provision of target assemblies, provision of an industrial site for development and commissioning of a demonstration set-up, system integration and control. In addition they are interested in the development of an optical sensing technique to allow dimensional checking to be integrated into the assembly process.

VUB are involved in the provision of a ver-satile gripper system, and with development of force and tactile sensing techniques. They are also looking at certain aspects of robot control and integration.

Risø and Dantec are working together on the development of a robot vision system. This will use optical rather than electronic image processing methods, and as such involves a degree of fundamental research. Broadly, Risø are covering the research while Dantec are developing a practical system.

The human factors areas are being addressed by SAPU at the University of Sheffield, working interactively with the system designers.

Additionally, Westland have so far employed two sub-contractors. Inbucon-Berric Ltd. have examined the design of a cell transport system, including selection of a conveyor system, the development of a control strategy, and design of a flexible transport pallet. Sira Ltd. have studied existing optical gauging methods, match-ing them to typical assembly metrology problems, and have recommended directions for further work.

4   CURRENT PROGRESS

4.1   Review Studies

The project started in January 1985. Much of the first year's work has been concerned with reviews of the 'state of the art' in various areas, with the primary aim of establishing a realistic level of ambition for the project. These studies have been accompanied by a process of defining and refining specifications and designs for the system and its various sub-areas. Additionally, much useful practical work has already been carried out.

Thus Westland, with assistance from VUB, have carried out a general review of flexible assembly systems, which includes an historical survey of automated assembly, an overview of flexible assembly, studies of each of the individual enabling technologies, examples of major research programmes, and examples of systems operating industrially.

In addition Westland have also sponsored two other surveys. One, carried out internally, looked at 'CAD system requirements for the off-line programming of an automated flexible assembly cell', and has laid the foundation for future project work on offline programming of assembly robots. The other, already mentioned, was carried out by Sira Ltd. and looked at optical metrology techniques applicable to flexible assembly.

Risø and Dantec have also carried out a review of robot vision, with the emphasis on optical image processing. Topics covered include system architectures, holographic optical elements, and spatial light modulators.

Sheffield have carried out reviews of human factors work in computer integrated manufacture and in manufacturing generally, but their results will not be published until the end of year 2.

4.2   System Design

Early studies based on analysis of typical assemblies indicated that a cell structure based on a loop of conveyor belt was most suitable. This was subsequently confirmed by independent studies at Bath University and at Inbucon-Berric. The design shown in Fig. 2 has evolved.

A major constraint here is the aerospace industry demand for traceability, which precludes the use of vibratory or other types of feeder issuing parts from a hopper. Thus an alternative method of transporting and presenting parts to work-stations had to be found.

The approach was to classify component parts to be transported in terms of stability and in terms of ease of grasping by a robot gripper. A novel form of transport pallet and associated modular support blocks, to permit transport of a wide variety of parts in approximately fixed positions and orientations, has been designed.

Available conveyor systems were examined and a control strategy for the movement of parts and pallets around the cell was established. The mode of operation is summarised as follows:-

A kit of parts for a given assembly is conveyed to the cell, with the parts arranged on a delivery tray. A vision-guided robot sorts the parts onto transport pallets, which have themselves been configured by a robot at another station, for transfer to the appropriate work-stations for preparatory processes.

The most sophisticated work-station will consist of two or more robots operating over a manipulating work table, in a sensor rich environment, for the complex or final automated operations. Close by will be a manual work-station for those operations which it is not viable to automate.

Base components will be attached to the work table by means of a flexible fixturing system, developed at the University of Bath with Westland (but not Esprit) funding. These fixtures will be prepared at a dedicated work-station.

Further work-stations will be simpler and devoted to preparatory operations such as cleaning or lubrication, or to simpler assembly operations such as 'heat and freeze' pressing.

The central loop of tracking is known as the express route. Arranged around it are a number of subsidiary loops, called regional routes, and these spurs lead to the various work-stations. These may be grouped logically around a regional route either by virtue of having common service requirements or according to frequently recurring sequences of operations.

Transport pallets, each having a unique binary identification code, to be read at critical junctions, are despatched to proceed around the cell on the express route. The host computer contains information on pallet contents. A 'wanted list' of which pallets are needed where, is held by a tracking micro. This is downloaded to a number of local plc's such that a pallet at the top of a work-station 'wanted list' is diverted onto the appropriate regional route, and from there to the work-station when free. The host computer also contains data files which are packages of information relating to each component or sub-assembly in terms of the requirements of the gripper, vision system, fixturing and so on.

The component or sub-assembly is then replaced on a pallet, either the same one or a different one, to visit other work-stations as required. All parts are eventually brought together at the final assembly area.

This broad outline of the current system concept has provided a useful framework into which the activities of the other participants must be integrated. Present Westland work includes simulation of the cell function.

4.3 Gripper Design

VUB have analysed the requirements of a flexible gripper system and are favouring a complex, highly flexible gripper rather than a series of interchangeable grippers. The main reason for this is the problem of connecting the actuation media and the sensory linkages through the wrist in a quick-change system.

A four-fingered gripper has been designed, with each finger having two degrees of freedom and also rotating around its base to allow fingers to oppose each other in pairs or to face a common centre-point. The fingers are also mounted on a traversing mechanism to extend the grip width.

Gripper actuation will be by pneumatics for the finger joint motions and by electric stepper motors for the translation and rotation motions.

Gripper construction has begun and tests have been carried out on control of stepper motors and of pneumatic actuators. A cartesian robot system has also been installed for use on gripper development work. A VME-FORCE computer system will be used for control of the cartesian robot and the gripper, using the FORTH language. There is no mass storage device at present and an Olivetti personal computer has been linked to the VME system to emulate a standard terminal.

4.4 Tactile Sensing

A fibre-optic based tactile sensing technique is under investigation at VUB. Claimed advantages are immunity to electro-magnetic interference, low loss due to transmission length, safety, and low weight. Tests have been carried out on coupling of fibres with each other and with light sources. This area of work is currently being re-evaluated and alternative techniques are also under consideration. Force and torque sensing will also be addressed.

4.5 Vision Sensing

Risø and Dantec are jointly developing the cell's vision system. Their approach is to by-pass the conventional digital electronic processing methods and to construct a system using optical analogue image processing. Such a system will give very much greater speed of operation but is as yet not proven in such an application.

A review of system architectures and of individual system elements has been carried out. An example of the favoured architecture is shown in Fig. 3.

A lens captures the image. A spatial light modulator (SLM) converts incoherent to coherent light, and an optical Fourier Transform is made of the input image. The image position is represented in the Fourier plane by a constant phase factor, so the processing of the Fourier Transform is independent of object position. It is, however, highly sensitive to size and orientation. It may be rendered size and rotation invariant by the introduction of a phase filter and another Fourier transforming lens which together carry out a mapping from rectangular to log-polar co-ordinates such that rotation and size of the object are converted to shifts in the new co-ordinate system.

The intensity of the object in the transformed state is processed by the second SLM and another Fourier Transform. In this Fourier plane orientation, size and position are given by a phase factor and object form by an amplitude distribution. By inserting a matched filter or collection of multiplexed matched filters into the Fourier plane the next transforming lens produces a point or collection of points on the final output detector, if the matching is successful. Point co-ordinates are determined by object rotation angle, size and position. To obtain all four parameters, two independent measurements must be made.

Much preliminary research work, sufficient to prove the validity of the concept, has already been carried out at Risø and Dantec.

Mapping filters have been produced by reducing an A4 size computer generated grid down to a thick phase hologram. These have been successfully demonstrated, although there is still further work to be done here. Matched holographic filters are also under preparation. Initially the system will use multiplexed matched filters, but ultimately electrically-addressable spatial light modulators may be used. At present SLM technology is not sufficiently mature.

Dantec have carried out a review of SLM's. Contact has been made with British Aerospace Dynamics Group who are producing optically-addressable SLM's and will act as a sub-contractor to provide Dantec with these devices. These will be used for the incoherent-to-coherent conversion and for the transformation of the geometrical transformation processor output to the correlator input.

At present SLM resolution is the major limitation to system performance necessitating a trade-off between number of objects viewed simultaneously and measurement accuracy. However, it is expected that the technology will develop sufficiently within the timescale of the project to permit the aims to be achieved. The use of the vision system for active, real-time guidance of robots during the assembly process is a more long-term objective, but is nevertheless feasible.

A portable demonstration system based on the concept described above is currently under construction.

4.6 Optical Gauging

It is found that many mechanical assemblies require the accurate checking of certain dimensions during the assembly process. This may be as a result of tolerance build up, for example necessitating selective assembly. Current technology would require that the part be removed from an automated assembly process and measured by a conventional co-ordinate measuring machine or by hand.

The philosophy here is to integrate the metrology capability into the robotic assembly process. This part of the programme is being sub-contracted by Westland to Sira Ltd. in the UK. They have carried out a review of potentially applicable non-contacting techniques, concentrating on optics. In fact it is found that only two techniques offer adequate resolution, these being triangulation and interferometry. The former is only suitable if it can be used close to the workpiece, while the latter requires a 'friendly' reflecting surface to operate from.

Thus a two-stage measuring process is envisaged whereby a compact triangulation sensor is attached rigidly to the robot, close to the end effector, to measure points on the workpiece surface. The sensor's position in space relative to a fixed frame of reference is monitored by an array of laser sensors, impinging on corner cube reflectors attached to the robot, which combine dynamic tracking with interferometry.

4.7 Human Factors

The project team believe that a major obstacle to the successful implementation of advanced manufacturing systems generally is failure to properly integrate humans into the system. This may be the case at any level from assembly work and maintenance, through programming and scheduling, up to supervisory and management functions.

It is typically found that new technology is designed with little or no regard for human factors, which are only considered at the implementation stage. This is usually too late since major decisions on design and operation which should have human factor inputs have already been made. This can lead to numerous problems in terms of efficiency of operation as well as human issues such as job satisfaction, industrial relations difficulties and retrospective training.

The approach in this project is to adopt a parallel design philosophy whereby the technical and human aspects are considered together throughout the design and implementation phases. This is currently being achieved through regular meetings between the Sheffield psychologists and the Westland designers.

The objectives of the human factors component of the project may thus be summarised as follows:

- to include in the design process consideration of the human factors aspects of CIM;

- to perform an evaluation of these aspects within the cell;

- to provide a set of generalisable design guidelines for use in other manufacturing environments.

Eight human factors tasks have been identified:

- examination of human factors aspects in manufacturing;

- specification of human factors guidelines;

- identification of conflicts;

- determination of methods of measurement;

- measurement of human factors in existing assembly systems;

- measurement of human factors in demonstration cell;

- measurement of human factors in final assembly system;

- specification of generalisable human factors guidelines.

Work to date has consisted of literature surveys, which will result in a report at the end of year 2, visits to other organisations, discussions with the system designers, and the generation of a preliminary set of human factors design guidelines.

Seven areas of research have been identified, although this list is not necessarily finalised. They are:

- system design process and allocation of function;

- hardware ergonomics;

- environmental ergonomics;

- safety and health;

- software ergonomics;

- job design;

- organisational aspects.

A number of design guidelines have been identified within each area, and discussion is currently being held on how they should be presented to a designer, both physically and in terms of a structure to render them more easily usable.

5   FUTURE ACTION

Having more or less completed the preliminary phase of reviews and planning, the project is now well into its second phase which involves the development of the individual technologies.

In the system design area some target assemblies from the Normalair-Garrett range have been identified and analysed. These are providing the basis for a cell work flow simulation using the HOCUS package, which is aiding the optimisation of cell design and operation, and for a study of robot offline programming using the RAPT package from Edinburgh University.

Part of the cell conveyor system is being constructed to act as a focus for the other areas of work.

A working gripper is being produced at VUB and tested on a demonstration system under construction there. Further work will be carried out on the tactile and force sensing techniques.

Dantec are completing their vision system demonstration set-up and will look further at the question of spatial light modulator availability. Risø will continue developing the optical image processing technology, and in particular the generation of holographic mapping and matched filters.

Sira Ltd., as sub-contractor to Westland, are carrying out a detailed feasibility study on the interferometric laser robot tracking concept.

Sheffield are completing their literature surveys and will continue to develop their design guidelines interactively with the system designers.

A major task for the Westland team is the integration of these contributory work packages. It has been recognised that the project would benefit from the addition of a sixth partner with particular experience of robot application in an industrial environment. This partner is currently being sought.

ACKNOWLEDGEMENTS

This paper is a compilation of summary reports by ESPRIT Project S34 partners. The authors also wish to acknowledge the continued support of colleagues in their respective companies and of friends and colleagues at Inbucon-Berric Ltd., Sira Ltd., the University of Bath, Imperial College, the ESPRIT Directorate and other ESPRIT contractors.

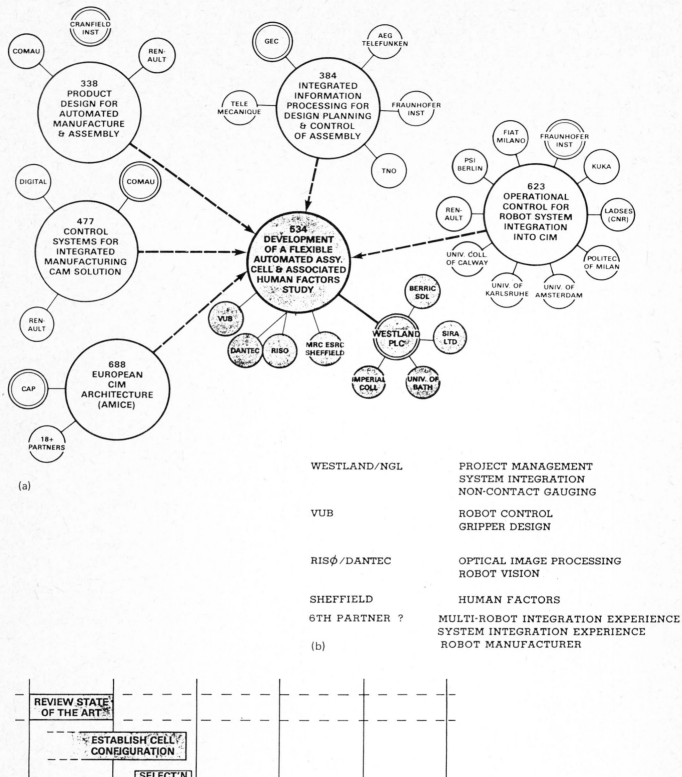

(a)

WESTLAND/NGL	PROJECT MANAGEMENT SYSTEM INTEGRATION NON-CONTACT GAUGING
VUB	ROBOT CONTROL GRIPPER DESIGN
RISØ/DANTEC	OPTICAL IMAGE PROCESSING ROBOT VISION
SHEFFIELD	HUMAN FACTORS
6TH PARTNER ?	MULTI-ROBOT INTEGRATION EXPERIENCE SYSTEM INTEGRATION EXPERIENCE ROBOT MANUFACTURER

(b)

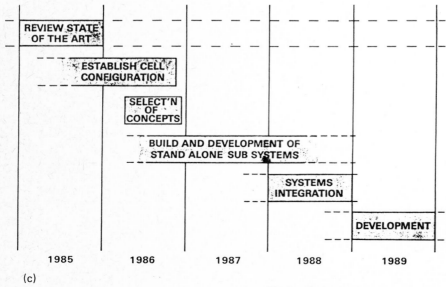

(c)

Fig 1   Project structure

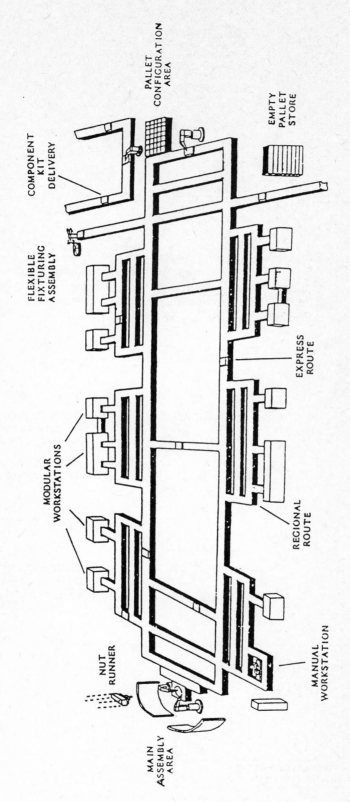

Fig 2 Current cell concept

PALLET CONFIGURATION AREA

COMPONENT KIT DELIVERY

EMPTY PALLET STORE

FLEXIBLE FIXTURING ASSEMBLY

EXPRESS ROUTE

MODULAR WORKSTATIONS

REGIONAL ROUTE

MANUAL WORKSTATION

NUT RUNNER

MAIN ASSEMBLY AREA

General Vision System

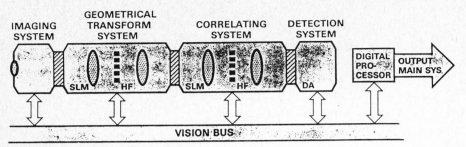

SLM: SPATIAL LIGHT MODULATOR
HF: HOLOGRAPHIC FILTER
DA: DETECTOR ARRAY

The transform system uses a computer generated filter to perform a cartesian to log-polar mapping in order to obtain rotation invariance. The correlating system uses a multi-matched filter to identify objects. The position of the correlation spot on the detector indicates the position and rotation of the object.

Fig 3   Vision system architecture